.

Numerical Modeling of Water Waves in Coastal and Ocean Engineering

Advances in Coastal and Ocean Engineering

ISSN: 1793-0731

Series Editor: Philip L-F Liu *(Cornell University, USA)*

Published

To view the complete list of the published volumes in the series, please visit
http://www.worldscientific.com/series/acoe

ADVANCES IN
COASTAL AND OCEAN ENGINEERING
VOLUME 13

Numerical Modeling of Water Waves in Coastal and Ocean Engineering

Pablo Higuera

National University of Singapore, Singapore
The University of Auckland, New Zealand

Jinghua Wang

The Hong Kong Polytechnic University, Hong Kong

Jie Hu

National University of Singapore, Singapore

Zhengtong Yang

Technology Centre for Offshore and Marine, Singapore

World Scientific

NEW JERSEY · LONDON · SINGAPORE · BEIJING · SHANGHAI · HONG KONG · TAIPEI · CHENNAI · TOKYO

Published by

World Scientific Publishing Co. Pte. Ltd.

5 Toh Tuck Link, Singapore 596224

USA office: 27 Warren Street, Suite 401-402, Hackensack, NJ 07601

UK office: 57 Shelton Street, Covent Garden, London WC2H 9HE

Library of Congress Cataloging-in-Publication Data
Names: Higuera, Pablo, author.
Title: Numerical modeling of water waves in coastal and ocean engineering / Pablo Higuera,
    National University of Singapore, Singapore, The University of Auckland, New Zealand,
    Jinghua Wang, The Hong Kong Polytechnic University, Hong Kong,
    Jie Hu, National University of Singapore, Singapore,
    Zhengtong Yang, Technology Centre for Offshore and Marine, Singapore.
Description: Singapore ; Hackensack, NJ : World Scientific, [2023] |
    Series: Advances in coastal and ocean engineering, 1793-0731 ; vol. 13 |
    Includes bibliographical references and index.
Identifiers: LCCN 2022047098 | ISBN 9789811265457 (hardcover) |
    ISBN 9789811265464 (ebook for institutions) | ISBN 9789811265471 (ebook for individuals)
Subjects: LCSH: Water waves--Mathematical models.
Classification: LCC TC172 .H54 2023 | DDC 627/.042015118--dc23/eng/20221202
LC record available at https://lccn.loc.gov/2022047098

British Library Cataloguing-in-Publication Data
A catalogue record for this book is available from the British Library.

For any available supplementary material, please visit
https://www.worldscientific.com/worldscibooks/10.1142/13118#t=suppl

Desk Editors: Soundararajan Raghuraman/Steven Patt

Typeset by Stallion Press
Email: enquiries@stallionpress.com

Foreword

In recent years, rapid growth of computing power has encouraged coastal and ocean engineers to use advanced numerical models to simulate and study complex water wave dynamics. Ocean waves are multi-dimensional and involve many physical processes. However, not every engineering problem entails all the physical processes. Therefore, many different wave models have been developed with their own approximations for a wide range of applications. Before using any such model, it is essential to understand the assumptions behind it and its limitations.

In this review volume, a comprehensive review of various water wave models is provided. Using time (length) scale of the wave processes to be resolved as a measure, wave models are grouped into phase-averaged models (Part 1) and phase-resolving models (Part 2). Within each group, models are categorized based on the different physical processes they can simulate, arranged from simpler to more complex. These include depth-averaged models (Chapter 3), potential flow models (Chapter 4), Navier–Stokes models (Chapter 5) and Lattice–Boltzmann models (Chapter 6). Each approach is reviewed in terms of its historical development, the underlying assumptions and limitations, the most relevant models and their practical applications. The methodologies for coupling among different models are reviewed in Part 3, which outlines the different types of implementations, highlighting their pros and cons, and provides relevant examples available in the literature.

The review volume is written for a wide audience, ranging from MSc and PhD students to research engineers and consultants. On the one hand, the complete and up-to-date list of references is useful for beginners to enter the field of wave modeling. On the other hand, the examples presented provide a practical way to select the most suitable approach to solve a certain problem. Readers are expected to have some basic knowledge of coastal processes (wave hydrodynamics, water wave theories and wave transformations), which are briefly outlined in Chapter 1.

Philip L.-F. Liu
July, 2022

About the Authors

Pablo Higuera is currently a Catastrophe Risk Modeler at Moody's RMS and an Honorary Academic at The University of Auckland. His past experience includes being Lecturer in Coastal Engineering at the University of Auckland, Senior Research Fellow at the National University of Singapore and Research Associate at Imperial College London. During his research career he has designed and developed the innovative numerical wave CFD modeling tool olaFlow, released as open source and used globally, with more than 120 published references.

Dr Higuera's main interests include numerical modeling of waves, Computational Fluid Dynamics (CFD), tsunami risk, coastal inundation and machine learning.

Jinghua Wang is an Assistant Professor in the Department of Civil and Environmental Engineering at The Hong Kong Polytechnic University. His research focuses on the mechanisms of nonlinear ocean disastrous waves, including rogue/freak waves, tsunamis and storm surges, etc., and their interactions with ocean and coastal engineering applications. His research interests are widely associated with developing numerical simulation tools with emphasis on addressing the unprecedented need for coastal resilience and sustainability.

Hu Jie is currently a hydrodynamic modeler in charge of the MetOcean model development in the coastal department of Surbana Jurong. Prior to joining Surbana Jurong, Dr. Hu Jie was a research fellow at NUS, where he obtained his Ph.D. degree in coastal engineering. Dr. Hu's research interests include both traditional breakwater and green infrastructures on the reduction of wave disasters, modeling simulation on inland/overland flooding and wind–wave interactions. Dr. Hu has been involved in the PUB project of coastal-inland flood model and contributed to the development of the coastal flood model as a part to better assess the impact of climate change on Singapore's coastal areas. Dr. Hu also had experiences in diverse research laboratories in France and several publications in the top journals in coastal engineering.

Yang Zhengtong was born in China in April, 1992. He obtained his Bachelor's and Master's degrees from the Ocean University of China in 2013 and 2016, respectively. Then he relocated to Singapore in July 2016 to continue his research in coastal engineering in the Department of Civil and Environmental Engineering at the National University of Singapore, where he was awarded the PhD degree in 2021. Immediately after completing his Ph.D. thesis, he started in the position of Scientist at the Technology Centre for Offshore and Marine, Singapore (TCOMS). His research interests are directed toward a better understanding of coastal processes, such as nearshore circulations, wave evolution from generation to the shoreline, especially theoretical and numerical modeling of wave transformation and wave-current interactions in nearshore regions.

Acknowledgments

This research was funded by PUB, Singapore's National Water Agency.

The authors would like to thank Prof. Philip L.-F. Liu for his guidance during the writing of this book.

Contents

List of Figures

List of Tables

Chapter 1

Introduction

Ocean waves are the main forces driving offshore and coastal processes. From wind-wave generation in the middle of the ocean to storm surge and surf and swash processes at the coast, the range of temporal and spatial scales associated with ocean waves are very wide.

In terms of wave period (or temporal scale), ocean waves can be classified as shown in Fig. 1.1. From capillary waves to wind waves, wave periods range from less than 1 second to tens of seconds. Wind-generated waves can impact offshore structures in deep water or travel up to thousands of kilometers, being filtered in the propagation process, and arrive at the coast as swells. Infragravity waves have periods ranging from tens to hundreds of seconds and are often bounded to propagating wave groups. Free infragravity waves are generated as the wave groups interact with coastal structures such as breakwaters or experience rapid changes of bathymetry and currents. Infragravity waves play a significant role in the run-up and inundation processes at the coast. Tsunamis are rare but powerful ocean waves with very long periods, in the range of minutes to hours, often generated by underwater earthquakes or submarine landslides. Despite their rare occurrence, their destructive impacts on the coast have attracted great attention of the coastal engineering community. Finally, storm surges and tides are even longer waves, with periods as long as hours or even days. Although the generation mechanisms

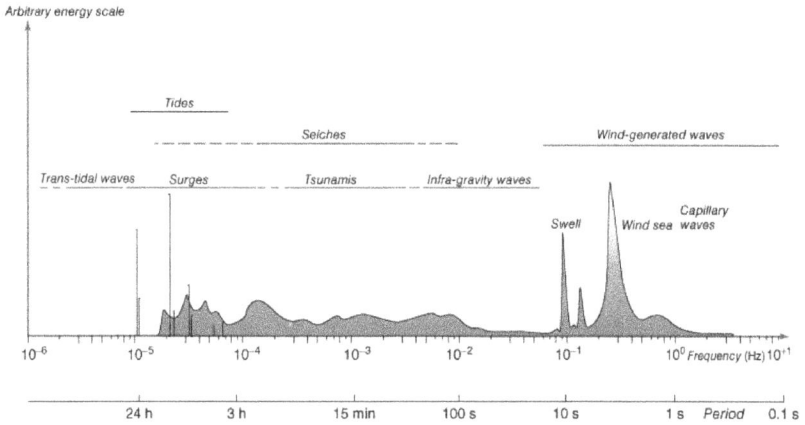

Fig. 1.1   Classification of ocean waves in terms of their period. Reproduced with permission from Holthuijsen (2010).

for tides (Moon and Sun attraction) and storm surges (atmospheric pressure depression) are very different, their effects on the low-lying coastal areas are very similar, namely sustained water levels that can flood the coast under certain circumstances.

The characteristic wave period is associated with the characteristic wavelength via the dispersion relationship (Dean and Dalrymple, 1991). For small amplitude waves propagating in a constant water depth, the longer the wave period is the longer the wavelength becomes. Consequently, the spatial scales of ocean waves are also vast, ranging from a few millimeters for capillary waves to thousands of kilometers for tides and storm surges.

At the same time, ocean waves interact with the environment, leading to physical processes occurring at a diverse range of scales. For example, ocean waves can propagate for thousands of kilometers across the ocean, interacting with various geological features (e.g., bays and headlands) at the scale of tens of kilometers or with coastal structures in the hundreds of meters to kilometers scale. As they approach the coast, ocean waves become more depth-limited and will break eventually as the water depth diminishes. Wave breaking is a dynamic process with the length scale of tens of meters, which in turn triggers significantly smaller processes, on the order of millimeters or smaller, such as the mobilization of sediment grains, air entrainment or turbulence. The ocean waves are eventually dissipated after breaking.

Apart from different temporal and spatial scales, the wave transformation processes also have different degrees of complexity. Whereas some processes can be explained by deterministic analytical expressions, semi-empirical formulations or direct numerical simulations (e.g., wave shoaling, wave refraction, wave reflection and wave scattering); others like wave breaking are so complex that they can only be approximated by simplified models. In this book the words *simulate* and *model* will be used with specific meanings. On the one hand, if the governing equations can "perfectly" describe a given process, the numerical solutions of the governing equations will be referred to as capable of *simulating* such a process. On the other hand, if a particular process cannot be described exactly mathematically, a simplified model must be introduced. The corresponding numerical solutions will be referred to as capable of *modeling* such a process.

Because of the diversity of scales and complexities, there is no single approach (neither physical experimentation nor numerical modeling) that can currently deal with all the scales/processes at once. Thus, this book shall review a number of different numerical approaches, each of which has its own range of applicability, advantages and disadvantages. To a certain extent, this review can be considered as an extended and up-to-date version of the "Wave propagation modeling in coastal engineering" paper by Liu and Losada (2002).

The main numerical modeling approaches to simulate ocean waves in the context of coastal and offshore engineering are included in Fig. 1.2. The first criterion divides the numerical models based on whether they study the wave evolution in terms of the energy of each frequency component (phase-averaging approach) or the evolution of each individual wave in time (phase-resolving approach). The phase-averaging approach is mostly limited to wave spectral models, which are very efficient in calculating local wave generation by wind and wave propagation over very long distances.

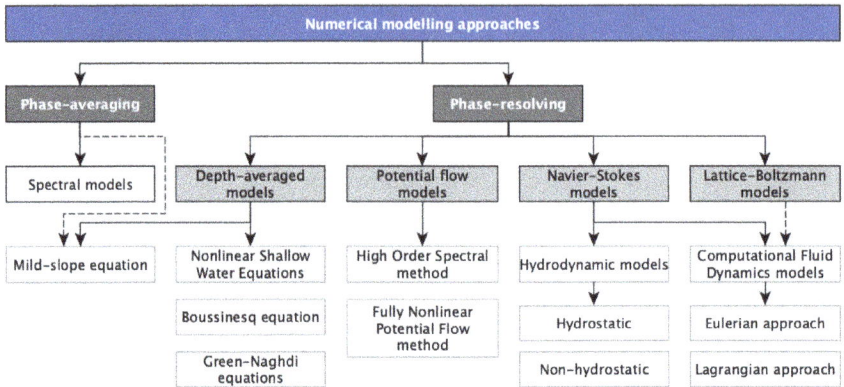

Fig. 1.2   Diagram of the most widely used numerical modeling approaches for coastal/offshore engineering.

In terms of the relative water depth (i.e., the ratio between the water depth and the wave length, h/L), ocean waves can be classified into three groups; namely, deep water regime for $h/L > 1/2$; shallow water regime for $h/L < 1/20$; intermediate water regime otherwise. In the deep water and shallow water regimes, ocean waves exhibit different characteristics, allowing further simplifications. Some of the subgroups in the phase-resolving group have been formulated using these simplifications as assumptions. For example, some depth-averaged models were originally formulated to represent waves in the shallow water regime. Nevertheless, their range of applicability has been extended by adding terms in the model equations. Potential flow models are formulated to model all relative water depths. However, similar to the depth-averaged models, the potential flow models are limited to non-breaking conditions. If wave breaking occurs, the process needs to be further modeled.

Numerical models based on Navier–Stokes (NS) equations can also represent ocean waves in all relative water depth regimes. Among these, hydrodynamic models solve simplified sets of NS equations, which limits the types of processes that they can simulate, but extends the range of applicability in terms of the spatial and temporal scales that they can deal with. Finally, numerical models based on the full NS equations (and by extension Lattice–Boltzmann models) are know as Computational Fluid Dynamic (CFD) models, and are the most advanced models available presently. CFD models can simulate all wave processes in detail, and they can even simulate wave breaking since the physical processes that it comprises are captured by the equations. Nevertheless, as discussed later, applying CFD models requires a very large computational cost. Therefore, the temporal and spatial domains simulated by this approach are heavily constrained by the computational resources available and the simulation time requirements.

Table 1.1 Notation used in this book in alphabetical order, unless otherwise noted in the respective sections.

A	Wave amplitude [m]
c	Wave celerity [m/s]
C_b	Bottom friction coefficient [m^2/s^2rad^2]
$\mathbf{c_g}$	Wave group celerity vector [m/s]
c_θ	Advection speed in the direction domain [rad/s]
c_σ	Advection speed in the frequency domain [rad/s^2]
D_x, D_y	Diffusion terms [m/s^2]
E	Wave energy density spectrum [m^2 s/rad]
f	Particle distribution function
f_c	Coriolis force [N]
$\mathbf{g}$	Gravitational acceleration vector [m/s^2]
H	Total water depth [m]
h	Water depth [m]
i	Imaginary unit
$\mathbf{k}$	Wave number vector [m^{-1}]
L	Wave length [m]
N	Wave action density [m^2 s^2/rad]
p	Total pressure [Pa]
p_a	Atmospheric pressure [Pa]
$\mathbf{r}$	Cartesian coordinates vector (x, y, z) [m]
$\mathbf{r_H}$	Cartesian horizontal coordinates vector (x, y) [m]
S	Spectral forcings [m^2 s /rad]
t	Time [s]
$\mathbf{U}$	Velocity vector (u, v, w) [m/s]
$\mathbf{U_H}$	Horizontal velocity vector (u, v) [m/s]
u, v, w	Velocity components [m/s]
$\bar{\mathbf{U}}$	Depth-averaged velocity vector $(\bar{u}, \bar{v})$ [m/s]
$\bar{u}, \bar{v}$	Depth-averaged velocity components [m/s]
x, y, z	Cartesian coordinates components [m]

Note: Bold letters indicate a vector; if used in regular font, they usually represent the modulus of the vector.

Table 1.2   Symbols used in this book in alphabetical order, unless otherwise noted in the respective sections.

α	Volume of Fluid indicator function
β	Biphase [rad]
γ	Interaction coefficient [s^{-1}]
ϵ	Wave steepness ratio (A/L)
ζ	Complex-valued amplitude of η
η	Water free surface elevation [m]
θ	Wave direction [rad]
λ	Depth to wavelength ratio (h/L)
μ	Fluid molecular dynamic viscosity [N s/m^2]
ν	Fluid molecular kinematic viscosity [m^2/s]
$\boldsymbol{\xi}$	Particle velocity vector [m/s]
ξ_x, ξ_y, ξ_z	Particle velocity components [m/s]
ρ	Fluid density [kg/m^3]
σ	Wave intrinsic circular/angular frequency [rad/s]
τ	Stress tensor [N/m^2]
φ	Wave phase [rad]
ϕ	Velocity potential [m^2/s]
ω	Wave Doppler-shifted circular/angular frequency [rad/s]
∇	Gradient vector (Hamilton operator)
$\nabla_{\mathbf{H}}$	Horizontal gradient vector
Δ	Laplacian operator

Note: Bold symbols indicate a vector.

Often times the limitations of one numerical model can be overcome by other models. Therefore, "connecting" different models seamlessly has a high value to solve problems that cannot be solved by one model alone. This approach can be used, for example, to extend the area covered by the simulation, to reduce the computational cost or simulation time, or even to add new physics in addition to hydrodynamics, such as solid mechanics or sediment transport. As a result, one of the last sections of this book is devoted to couplings between numerical models.

This book is structured as follows. The index of variables used in the book is included at the end of this introductory chapter, in Tables 1.1 and 1.2. Then, the phase-averaged wave simulation approach is described, and spectral wave models are introduced in Part 1, which comprises Chapter 2 alone. Next, the phase-resolving approaches are presented in Part 2, which starts with Chapter 3. This part is the most extensive, as it includes four modeling groups

(Depth-averaged models, Potential flow models, Navier–Stokes models and Lattice–Boltzmann models), presented in increasing order of complexity. Each of these modeling groups is presented in its own chapter, and may include several modeling approaches (in terms of the equations solved). Then, Chapter 7 reviews the main techniques to couple different wave modeling approaches and additional models, discussing the advantages and complexities of the different types of couplings. Finally, the book ends with an executive summary, which highlights the most important conclusions of this book, and an appendix, in which the most relevant numerical models available in the literature are listed.

Part 1

Phase-Averaged Approaches

The phase-averaged approach describes the evolution of random wave spectra in space and time accounting for the wave generation, propagation and dissipation. In this approach, the amplitudes of the frequency (or wavenumber) modes are deterministic and the phases are distributed randomly. Because of this, phase-averaged models most often lead to wave field solutions that can provide the wave spectral shape and wave statistics such as the significant wave height or mean wave period at a given location and time, but not the time history of the instantaneous wave surface elevation or velocity information. Nevertheless, assuming that the wave phases are randomly distributed (see Section 7.1 for further information), the wave surface elevation can be reconstructed indirectly by linear superposition of all frequency (or wavenumber) modes. This modeling approach is also called spectral wave modeling, which will be reviewed next.

Chapter 2

Spectral Wave Models

2.1 Spectral wave models

2.1.1 *Introduction*

Spectral models are intended to reproduce how waves are generated by wind and how they evolve as they propagate toward the coast. This is why spectral models are also known as operational models, as they can inform the operations of, for example, harbors by providing a forecast of the wave conditions days in advance.

Spectral models are characterized as first-, second- or third-generation models based on their capabilities to deal with nonlinear wave–wave interactions, as explained in Caetano and Innocentini (2003):

- In the first-generation models, the wave spectrum evolves until it reaches the saturation level formulated in Phillips (1957) and the components do not interact nonlinearly with each other, which is why first-generation models are also called decoupled models.
- In the second-generation models, the nonlinear wave–wave interactions are parameterized assuming a given spectral shape beforehand, which is why they are also called parametric models. Usually, the JONSWAP spectrum (Hasselmann *et al.*, 1973) is used as the target spectral shape for the pre-computed wave–wave interactions.
- In the third-generation models, the nonlinear wave–wave interactions are computed explicitly, therefore, "the spectrum is free to

develop without any shape imposed *a priori*" (Holthuijsen, 2010). These are the most advanced spectral wave models and will be reviewed next.

2.1.2 *Governing equations*

In random seas, the free surface elevation (η) can be generally written as the summation of n frequency (or wavenumber) modes as follows:

$$\eta(\mathbf{r_H}, t) = \sum_{j=1}^{n} A_j \cos(\mathbf{k}_j \cdot \mathbf{r_H} - \sigma_j t + \varphi_j), \qquad (2.1)$$

where A_j is the amplitude of jth mode and can be obtained by the expression $A_j = \sqrt{2E(\sigma, \theta)\Delta\sigma\Delta\theta}$, in which $E(\sigma, \theta)$ is the directional wave energy density spectrum, $\Delta\sigma$ and $\Delta\theta$ are the frequency and direction bin size, respectively. The wavenumber magnitude $k_j = |\mathbf{k}_j|$ and angular frequency σ_j are related by the dispersion relation of linear waves as follows:

$$\sigma_j^2 = g\, k_j \tanh(k_i\, h), \qquad (2.2)$$

where g is the acceleration due to gravity and h is the local water depth.

In Eq. (2.2), the relative phase of the component φ_j is often taken as random, in the absence of additional information on the free surface elevation. Therefore, the sequence $\{\varphi_j, j = 1, 2, \ldots, n\}$ is randomly and uniformly distributed in $[0, 2\pi)$.

Retaining only the information of the wave amplitude, the governing equation for the spectral wave model is the wave action balance equation (Komen *et al.*, 1996):

$$\frac{\partial N}{\partial t} + \nabla_{\mathbf{H}} \cdot [(\mathbf{c_g} + \mathbf{U_A})N] + \frac{\partial}{\partial \sigma}(c_\sigma N) + \frac{\partial}{\partial \theta}(c_\theta N) = S_{\text{tot}}, \qquad (2.3)$$

where $N = E(\sigma, \theta)/\sigma$ is the wave action density, $\nabla_{\mathbf{H}}$ is the horizontal gradient operator, $\mathbf{c_g}$ is the wave group velocity, $\mathbf{U_A}$ is the advection velocity depending on the current profile and wave amplitude, c_σ and c_θ are the advection speed in frequency and direction domain, respectively, and S_{tot} represents the source terms. The left-hand side is the kinematic part. The first term represents the rate of change of wave action density, the second term denotes the propagation of wave energy in the geographical space, the third term indicates the frequency shift due to the variation of water depth and the presence of current, and the fourth term describes the depth- and current-induced refraction.

The right-hand side represents the forcing by all physical processes including generation, dissipation and redistribution of wave energy. In operational models, this term is usually expressed as a summation of the nonlinear processes as follows:

$$S_{\text{tot}} = S_{\text{in}} + S_{\text{nl4}} + S_{\text{nl3}} + S_{\text{wc}} + S_{\text{br}} + S_{\text{bf}} + S_{\text{bg}} + \cdots , \qquad (2.4)$$

where S_{in} denotes the energy transfer from wind to waves, S_{nl4} and S_{nl3} represent nonlinear four-wave (quadruplet) and three-wave (triad) interactions, respectively, S_{wc}, S_{br} and S_{bf} are dissipation terms due to whitecapping, depth-induced breaking and bottom friction, respectively, and S_{bg} stands for the Bragg scattering due to wavy seabed effects. The relative importance of the relevant processes is summarized in Table 2.1. The historical development of spectral wave model is fully documented in Komen *et al.* (1996) and

Table 2.1   Relative importance of the various processes affecting the evolution of waves in oceanic and coastal waters.

Process	Oceanic waters	Shelf seas	Coastal waters Nearshore	Harbor
Wind-wave generation	●●●	●●●	●	○
Quadruplet resonance	●●●	●●●	●	○
Whitecap	●●●	●●●	●	○
Bottom friction	○	●●	●●	○
Current refraction/energy bunching	○/●	●	●●	○
Bottom refraction/shoaling	○	●●	●●●	●●
Depth-induced breaking	○	●	●●●	○
Triad resonance	○	○	●●	●
Reflection	○	○	●/●●	●●●
Diffraction	○	○	●	●●●

Note: ●●● = dominant, ●● = significant but not dominant, ● = minor importance, ○ = negligible.
Source: Reproduced with permissions from Holthuijsen (2010).

Cavaleri *et al.* (2007, 2018). Nevertheless, some key progress will be summarized in what follows.

2.1.3   *Wave propagation*

The left-hand side of the wave action equation describes the wave propagation including the effects of refraction, shoaling, diffraction and reflection, which are caused by the spatial variation of water depth and current. The equations for the spectral advection speed c_σ and c_θ in Eq. (2.3) are usually called "ray equations" and control the changes of wave propagation direction (refraction) and wave length (shoaling). For natural topographies, this approximation is quite robust. On the scale of several wave lengths, the representation of the refraction–diffraction effects was introduced to spectral models by Holthuijsen *et al.* (2003). Wave reflection at the shoreline can be treated by using a proper boundary condition for Eq. (2.3). Additionally, the effects of a wavy seabed are usually represented by a Bragg scattering source term on the right-hand side of Eq. (2.3) (Ardhuin, 2001; Ardhuin and Herbers, 2002; Ardhuin and Magne, 2007).

For waves propagating over a current, the current velocity is assumed to be uniform over the depth in spectral models. Currents can compress/stretch the wave length, change the wave group velocity and wave height. A spatially varying current can also refract the waves causing local focusing of wave energy in space. In addition, when the opposing current is sufficiently strong, wave propagation can be blocked. In practical wave forecasting, the tidal currents are often taken into consideration and are treated as steady currents. While some spectral models also consider the effects of unsteady currents, it has been discovered that the change of absolute frequency due to time variations in the current is of the same order of magnitude as the Doppler shift (Tolman, 1990, 1991). However, there is very little validation of wave propagation over horizontally varying currents. Recently, it has been found that the changes of spectral properties due to fully nonlinear interactions between waves and horizontally varying currents can be significantly different from the predictions based on the wave action equation (Wang *et al.*, 2021). Besides, the wave action equation for depth-varying currents has not yet been developed (Cavaleri *et al.*, 2007).

2.1.4 Wind–wave generation

The primary process by which waves are generated and grow is the transfer of energy between the wind and the surface of the sea. There are two fundamental concepts:

- the sheltering theory by Jeffreys (1925, 1926) and
- the resonance theories by Phillips (1957) and Miles (1957).

Jeffreys's theory assumed that air flow separation occurred at the leeward side of the wave crest, leading to a pressure difference, so that work could be done by wind on waves. Laboratory experiments showed that this sheltering mechanism yielded a pressure difference that was too small to drive the observed growth rate. Moreover, the separation only occurred over steep waves, so it must be applied locally in space and time rather than uniformly and constantly on the wave field (Banner and Melville, 1976). Therefore, Jeffreys's sheltering theory was abandoned and was not adopted in spectral models.

Phillips's resonance mechanism considered the resonant forcing of surface waves by turbulent pressure fluctuations, which produces a linear growth rate of the spectral energy. Meanwhile, Miles's feedback mechanism considered the resonant interaction between the wave-induced pressure fluctuation and the free surface waves, which yields an exponential growth rate. The two types of resonance mechanisms play important roles at different stages of wave growth. When the sea surface is nearly flat (wave phase celerity or wave age are small), pressure fluctuations induced by the air turbulence and traveling with the average wind speed will transfer the momentum to the sea surface disturbances according to Phillips's resonance theory. After this initial growth stage, the air flow will develop higher and lower pressure zones at the windward and leeward side of the wave crests, respectively, creating a downwind thrust. Since the motion of the water waves is in phase with the air pressure distribution, the process is in resonance and waves are amplified. Therefore, this stage is dominated by the Miles mechanism, and the growth rate is several orders greater than the initial stage of the Phillips mechanism. The long and short (or fast- and slowly-moving) waves also experience different growth rates determined by their wave age. In general, the short (slow-moving) waves are subjected to a greater growth rate than long (fast-moving) waves. Therefore, the short waves will become saturated first, followed by the long waves.

On a later stage and on a much longer time scale than wind-generated waves, the wind variability (gustiness) must be considered (Janssen, 1986). This will affect the waves with phase celerity close to the mean wind speed the most, as the fluctuation of wind speed may have considerable impact on wave growth (so-called "diode" effects, Abdalla and Cavaleri, 2002). For waves traveling with a phase celerity twice faster than the wind speed or waves propagating opposite to the wind direction, their energy may be transferred back (lost) to the atmosphere. As a result, those waves will be substantially damped. However, this mechanism is not well-understood so far (Cavaleri *et al.*, 2007).

On the other hand, the wind speed profiles (vertical distribution) are expected to change with the evolving ocean waves, and the energy transfer from the air to the waves may be affected by the sea state. To account for this effect, the quasi-linear theory of wind–wave generation was introduced (Janssen, 1982). The wave growth follows Miles's theory. However, Miles's constant becomes a function of the sea surface roughness length, that depends on wave-induced stress as a function of the wave spectrum (Janssen, 1989, 1992). Nevertheless, there are still other nonlinear effects that are not well-understood, such as the wave age dependence on the spectrum and sea-state dependence on the drag over waves (Makin and Kudryavtsev, 2002; Makin *et al.*, 1995).

In operational models, the wind–wave generation source term is usually expressed as

$$S_{\mathrm{in}}(\sigma,\theta) = A + BN(\sigma,\theta), \tag{2.5}$$

where A is the linear growth rate due to Phillips's mechanism, and B denotes the exponential growth due to Miles's mechanism. The latter is a function of wave age, which is calculated as the wave phase celerity divided by the friction velocity of air flow at the water surface. However, the friction velocity is usually not available, so the wind speed at 10 m height above the sea surface is employed in practice. A drag coefficient is often introduced to establish the relationship between the two quantities given a prescribed wind speed profile (vertical distribution).

2.1.5 *Nonlinear four-wave interactions (quadruplet resonance)*

After the emergence of wind waves, another process that affects the evolution of the wavefield is the wave–wave interactions, which transfer energy between different modes by resonance. Quadruplet wave–wave interactions are very important in deep and intermediate waters, where they redistribute the energy from the peak of the spectrum to lower and higher frequencies. In practice, quadruplets shift the spectral peak to lower frequencies, as higher harmonics increase the steepness of the waves, which eventually break (whitecapping) (SWAN Team, 2020). This phenomenon of energy cascade to higher frequencies is similar to the one observed in hydrodynamic turbulence, so it is also known as weak wave turbulence (Lavrenov *et al.*, 2001; Newell *et al.*, 2001). It is found that a set of four waves (quadruplet) can exchange energy when the resonance conditions are met as follows:

$$\mathbf{k}_1 + \mathbf{k}_2 = \mathbf{k}_3 + \mathbf{k}_4 \text{ and } \sigma_1 + \sigma_2 = \sigma_3 + \sigma_4, \tag{2.6}$$

where $\mathbf{k}_i$ and σ_i are the wavenumber and angular frequency of the ith component, respectively.

The theory to describe and reproduce nonlinear quadruplet wave–wave interactions was originally developed by Hasselmann (1962) as an integral expression. This formulation is known as the Boltzmann integral. Later, Zakharov (1968) derived it in a form known as the kinetic equation. Using the Zakharov equation, Janssen and Onorato (2007) found that the nonlinear transfer rate becomes very small in intermediate water depths for $kh < 1.363$. This is also confirmed by an analysis based on the full Boltzmann equation (Van Vledder, 2001).

The Boltzmann integral requires a very high computational cost to solve due to its high complexity. In order to include this physical process in numerical models efficiently, the Discrete Interaction Approximation (DIA) was developed (Hasselmann and Hasselmann, 1985; Hasselmann *et al.*, 1985). This technique can reduce the computational cost significantly while "preserving the basic characteristics of the full solution" (Van Vledder, 2001). Despite its advantages, the DIA has noticeable shortcomings too, which have led to the development of more sophisticated techniques. Some of them are reviewed in

Van Vledder (2001). Moreover, recent works, such as Ponce de Leon and Osborne (2020), suggest that using the full Boltzmann interactions directly, instead of the DIA and its derivatives, may lead to spectral shapes that can explain the observation of a very high central peak and the "bimodality" phenomenon during Draupner storm.

2.1.6   *Nonlinear three-wave interactions (triad resonance)*

When the waves propagate from deep/intermediate water to shallow water, they become non-dispersive (all wave modes travel with the same phase celerity). As a result, the other type of nonlinear wave–wave interactions, i.e., the so-called triads or triplets, namely interactions between three waves, will dominate in shallow water under the resonance conditions as follows:

$$\mathbf{k}_1 + \mathbf{k}_2 = \mathbf{k}_3 \text{ and } \sigma_1 + \sigma_2 = \sigma_3. \tag{2.7}$$

Besides, this resonance condition can also be easily satisfied among the dominating mode $(\sigma, \mathbf{k})$ and bound harmonics $(2\sigma, 2\mathbf{k})$ and $(3\sigma, 3\mathbf{k})$ of a periodic wave train. The bound waves do not obey the dispersion relation in deep and intermediate water depths as they are bounded to the dominating mode (meaning they all travel with the same phase celerity), but they will grow rapidly in shallow water. Stokes-like waves (sharper crest and flatter trough) are created and they pitch forwards when propagating toward even shallower water depth. This process is called near-resonant triad interactions.

The triad-wave interactions play an important role in transferring energy from incident wave modes to higher-(super-harmonics) and lower-(sub-harmonics) frequency modes. The magnitude of energy transfer depends on the phase differences of the triplets, which are quantified with the so-called biphase as follows:

$$\beta = \varphi_1 + \varphi_2 - \varphi_3, \tag{2.8}$$

where φ_1, φ_2 and φ_3 are the phases of the triplets. Therefore, properly determining triad-wave interactions requires solving the coupled evolution model for both biphase and the spectrum. This can be derived from the Boussinesq equation for one-dimensional (long-crested waves) case (Herbers and Burton, 1997). Note that the development of Boussinesq equation will be discussed separately in Section 3.3.

In operational models, the biphase is approximated without the biphase evolution equation. Instead, it is estimated from the spectrum and the local water depth, which is known as the Lumped Triad Approximation (LTA) method (Eldeberky, 1996). This method uses the biphase at the peak frequency that is originally derived for self–self interaction. LTA leads to constant energy transfer from component of frequency σ to that of frequency 2σ, but not to subharmonics. Hence, the shortcoming of using LTA is that it tends to overestimate the energy of super-harmonics. Though the LTA cannot describe the energy cascade to the sub-harmonics, such low-frequency components are often treated as long waves bounded to the wave groups. Therefore, they can be recovered from Hasselmann's equilibrium theory (Hasselmann, 1962) or through the concept of "radiation stresses" (Longuet-Higgins and Stewart, 1962). Some improvements on modeling the sub-harmonics (or the so-called infra-gravity waves) in the spectral wave models as free waves have been made recently (Ardhuin *et al.*, 2014; Reniers and Zijlema, 2022; Rijnsdorp *et al.*, 2021).

2.1.7 *Dissipation — Wave breaking in deep water*

The physics of wave dissipation is not well-understood from the spectral perspective. Hence, the dissipation rate is difficult to be parameterized. In general, the spectral dissipation due to wave breaking in deep water can be categorized into four models as follows (Komen *et al.*, 1996; Polnikov, 1993):

- *Whitecap model:* This is the most widely utilized model in operational practice. It assumes that well-developed whitecaps, situated on the forward faces of breaking waves, exert downward pressure, producing negative work on the waves (Hasselmann, 1974).
- *Quasi-saturated model:* This model assumes that the wind input, wave-wave interactions and dissipation become balanced in the range where the spectrum reaches saturation. Any excessive energy is interpreted as dissipation (Phillips, 1985).
- *Probability model:* This model uses a probability model to predict the appearance of waves that are higher than the breaking limit or

travel faster than the limiting acceleration at the crest (Longuet-Higgins, 1969; Phillips and Hasselmann, 2012).

• *Turbulent model:* It assumes that the energy dissipation is induced by the turbulence in the water after wave breaking, and the dissipation is governed by the effective turbulent (eddy) viscosity (Polnikov, 1993).

Each model leads to different dissipation rates, e.g., from being a linear to quadratic function of the wave spectrum. These models strongly disagree with some experimental observations, and even experimental data are often in conflict. Cavaleri *et al.* (2007) concluded that "such a state of knowledge of physics of the wave breaking losses does not help modeling the wave dissipation." Therefore, this source term describing the energy dissipation in deep water is the most debated term.

In the operational models, the whitecap model is widely adopted to describe the dissipation in deep water. This was firstly introduced by Komen *et al.* (1984) based on Hasselmann's (1974) quasi-linear theory. Its general form can be expressed as

$$S_{\mathrm{wc}}(\sigma, \theta) = \gamma N(\sigma, \theta), \tag{2.9}$$

where γ is the interaction coefficient that can be formulated separately under different assumptions. This dissipation function is often used as a "closing" term fitted to the DIA and the wind input term in order to describe the well-known growth relationships. It should be noted that the mean (or peak) period is often underestimated by using SWAN with this dissipation model (Rogers *et al.*, 2003). Hence, other alternative forms of the whitecap model are also available in operational practice. For example, Alves and Banner (2003) derived a dissipation function accounting for wave breaking due to wave group modulation. The resulting formulation is local in the frequency spectrum, in contrast to the expression by Komen *et al.* (1984). Alves and Banner (2003) also considered the threshold behavior of the wave breaking by incorporating a saturation level of the spectrum. Based on this work, several other dissipation functions have been proposed, such as those by Makin and Stam (2003) and van der Westhuysen *et al.* (2007), as used in the WWMIII and SWAN CycleIII models with corresponding wind input functions, respectively. Some studies also focused on estimating the dissipation rate from field measurements (Babanin and Young, 2005; Manasseh *et al.*, 2006; Young and Babanin, 2006). For instance, Young and Babanin (2006) suggested a dissipation function incorporating the threshold of breaking onset, directional spreading of the dissipation distribution and the cumulative effects of long waves on short waves (i.e., straining effects) (Donelan, 2001). However, its closure with the DIA and the wind input term needs further research.

2.1.8 *Dissipation — Wave breaking in shallow water*

The breaking mechanism in shallow water is quite different from that in deep water. When waves approach the shoreline, the shoaling effects will enhance the wave height, leading to wave breaking when the ratio of wave height over water depth exceeds a certain limit. This depth-induced breaking in the surf zone is responsible for the energy dissipation. In operational models, the energy loss due to this type of breaking is modeled in analogy with the dissipation in a bore (Battjes and Janssen, 1978). Eldeberky and Battjes (1995) extended Battjes and Janssen's (1978) model to be used in spectral models, usually expressed as

$$S_{\mathrm{br}}(\sigma, \theta) = \frac{D_{\mathrm{tot}}}{E_{\mathrm{tot}}} N(\sigma, \theta), \qquad (2.10)$$

where D_{tot} and E_{tot} are the total dissipation energy and total spectrum energy, respectively. D_{tot} depends on the fraction of breaking waves (Booij *et al.*, 1999) and the maximum wave height at a given depth.

An alternative to the bore-based model is to rely on the probability density of breaking waves (Thornton and Guza, 1983). Based on field observations of wave height statistics in the surf zone, it is suggested that D_{tot} is a function of the probability density function (such as the Rayleigh distribution) of breaking waves times the fraction of breakers.

2.1.9 *Dissipation — Bottom friction in shallow water*

In addition to depth-induced wave breaking, another type of energy dissipation in shallow water is bottom friction. This effect is usually important and dominant in continental shelf seas with sandy bottoms. The dissipation rate can be written in terms of the ensemble average of the product of the bottom shear stress and the near bed wave orbital velocity. Much work has been done to study the bottom boundary layer structure in order to obtain these two quantities. The work on deriving the dissipation term due to bottom friction used in the spectral model has been summarized in Luo and Monbaliu (1994). In operational models, the empirical JONSWAP spectrum (Hasselmann *et al.*, 1973), the drag law model of Collins (1972) and the eddy-viscosity model of Madsen *et al.* (1988) are usually selected. The resulting expression can be written as

$$S_{\text{bf}}(\sigma, \theta) = -C_{\text{b}} \frac{\sigma^2}{g^2 \text{sinh}^2 kh} N(\sigma, \theta), \tag{2.11}$$

where C_{b} is the bottom friction coefficient. The most used value for C_{b} is originally derived for the JONSWAP spectrum (Hasselmann *et al.*, 1973), but Zijlema *et al.* (2012) shows that it can be used for both swell and wind–sea conditions. Nevertheless, it should be noted that this term does not consider the effects of wave–current interactions. Conceptually, the bottom stress will be enhanced due to strong interactions between waves and currents. However, there is currently a lack of supporting evidence from field observations.

There are also other studies on the bottom friction considering effects of wave-generated bedforms (ripples) (Tolman, 1994), which play an important role on characterizing the bottom roughness. Additional bottom dissipation mechanisms due to percolation and water–mud interactions are also under investigation, e.g., Winterwerp *et al.* (2007) formulated the dissipation source term considering the soft muddy bottom.

2.1.10   *Numerical models and applications*

The development of the spectral models has undergone three stages. The first-generation models considered the wind input and dissipation by whitecaps, and the second-generation models additionally considered the nonlinear four-wave interactions (Komen *et al.*, 1996). The state-of-the-art formulations are the third-generation models, which consider all relevant processes for the evolution of the spectrum explicitly, without prior restrictions. The development of the DIA for approximating the Boltzmann integral (Hasselmann and Hasselmann, 1985) and the formulations of whitecaps to close the problem in deep water (Komen *et al.*, 1984) triggered the development of the third-generation wave models, such as WAM[1] (Hersbach and Janssen, 1999; Komen *et al.*, 1996; WAMDI Team, 1988), WAVEWATCHIII[2] (WW3) (Tolman, 1991; Tolman *et al.*, 2009), TOMAWAC[3] (Benoit *et al.*, 1997), etc. In shallow water, the formulations for deep water processes can be adjusted, and shallow water effects such as bottom dissipation, depth-induced breaking, triad-wave interactions are supplemented. This leads to the development of SWAN[4] (Booij *et al.*, 1999), CREST (Ardhuin *et al.*, 2001), etc.

WAM, WW3 and SWAN have become the mainstream of the third-generation models. They discretize the wave action equation in physical space on structured grids and solve the equation using the Finite Difference Method (FDM). At oceanic scales, WAM and WW3 can be employed with higher-order schemes for advection in space, which is more efficient than the SWAN model adopting a second-order implicit propagation scheme (Cavaleri *et al.*, 2007).

The application of structured grids in coastal environments is problematic, because the sea state changes due to bathymetry on a smaller spatial scale as the waves approach the coast. This problem is usually treated by adopting the "nesting" grid approach, e.g., as used in WAM (Lahoz and Albiach, 1997).

More recently, spectral wave models have been developed to work on unstructured grids, such as TOMAWAC or CREST. However, they employ the method of wave characteristics and calculate the energy advection along ray trajectories. This Lagrangian method also becomes restrictive in cases with cyclonic wind fields and with the presence of strong tidal currents. This has motivated the development of the spectral model on unstructured grids, such as WWMIII based on the Finite Element Method (Hsu *et al.*, 2005; Roland, 2008) and the unstructured versions of SWAN based on Finite Volume Method (Qi *et al.*, 2009) and FDM (Zijlema, 2010).

---

[1]https://mywave.github.io/WAM/
[2]https://polar.ncep.noaa.gov/waves/wavewatch/
[3]http://wiki.opentelemac.org/doku.php?id=user_manual_tomawac
[4]https://swanmodel.sourceforge.io/

To address the importance of the wave–circulation interactions, the spectral wave models are usually coupled with the circulation models to improve the forecasting capability. A comprehensive review on the circulation models based on the nonlinear shallow water equations will be given in Section 3.2 and hydrostatic wave models in Section 5.1. More recently, attempts were made to couple operational atmosphere and wave modeling (Breivik *et al.*, 2015; Janssen and Janssen, 2004). Moreover, coupling between global/regional-scale phase-averaged spectral wave models and the local-scale phase-resolved wave models can further extend the range of applicability for the problems of interests, as will be discussed in Section 7.1.

2.1.11   *Limitations*

The limitations of spectral wave models are openly discussed in Cavaleri *et al.* (2007, 2018). Only the key issues are discussed here.

By design, spectral models cannot simulate wave reflection and diffraction, as they are connected to the wave phase (which is not accounted for in this type of models). However, there are formulations that relax these restrictions and allow simulating wave diffraction reasonably well (Holthuijsen *et al.*, 2003) or even the complete wave transformation processes (shoaling, refraction, diffraction, reflection and breaking) (Mase *et al.*, 2005) on a scale larger than the wave length. Nevertheless, the diffraction of waves at the scale of the wave length cannot be described by the spectral models, and it often requires a phase-resolving model.

The operational models relate the linear wave problem with a vertically averaged current. The wave action equation for surface gravity waves in the presence of vertically-structured current of arbitrary profile is given by Quinn *et al.* (2017). However, it is yet to be implemented in the operational models. Moreover, the present spectral models do not consider the quadratic nonlinearities of wave–current interactions and cannot account for the higher-order nonlinear effects due to horizontally shearing currents (Cavaleri *et al.*, 2018). When waves propagate from a current-free zone into a region with (obliquely) opposed current, the fully nonlinear wave–current interactions can trigger the quasi-resonant interactions, leading to the broadening of the spectral bandwidth and the directional spreading, a mitigation of the spectral peak downshift as well as an enhancement of the significant wave height (Wang *et al.*, 2021). More comprehensive studies are still required to verify and improve the spectral wave models to account for the nonlinear effects of the horizontally shearing current.

Although using the drag coefficient to establish the relationship between wind friction velocity at wave surface to the wind speed at 10 m height above sea surface is widely accepted, its value under hurricane/typhoon conditions may not be suitable. Besides, for the damping of long (fast moving) waves and waves moving against the wind, there is no reliable and sufficient observation data, hence quantification of the wave damping in the field still remains a challenge. In addition, the "diode" effects due to gustiness are not considered in the operational models, and a full quantification of its actual relevance is still missing.

For describing the wave–wave interactions, the Boltzmann integral was originally derived for deep water and later extended for shallow water while assuming a flat bottom. Hence, its validity for shallow water condition and sloping bottom needs to be verified. Besides, developing novel computational methods that are accurate and efficient to evaluate the Boltzmann integral in operational models is still a challenge.

To model the infragravity waves in spectral wave model, a local equilibrium is often assumed between the incident sea-swell waves and the bound infragravity waves. However, this local equilibrium approximation may result in an overestimation or underestimation of the incident infragravity waves depending on the coastal configuration. Therefore, a nonlinear source term is required to describe the energy transfer from incident waves to infragravity waves. Nevertheless, the use of LTA for expediting the computations of the triad interaction term only considers the self-interactions of the co-linear waves and the energy transfer to super-harmonics. Despite that the radiation stress concept is shown to be extendable for this purpose at an alongshore uniform coast (Reniers and Zijlema, 2022), such a source term for general coastal configurations is not available yet, and new theories and methodologies are still demanded.

Though energy dissipation due to wave–turbulence interactions is an important wave energy sink, a consistent way of parameterization in the operational wave models still needs to be found. Furthermore, modeling the dissipation based on exact physics would be desirable in the future.

Part 2

Phase-Resolving Approaches

The phase-resolving approach solves the generation, propagation and dissipation of waves in the time domain. As a result, phase-resolving models can describe the evolution of individual waves at different stages during a simulation, being able to provide time series of the relevant variables (free surface elevation, velocities, pressure, etc.) In this section, 10 types of models grouped into the four categories already defined in the introductory chapter (depth-averaged models, potential flow models, Navier–Stokes models and Lattice–Boltzmann models) will be reviewed.

Chapter 3

Depth-Averaged Wave Models

Waves and currents are two dominating flow features in coastal waters, comprising various physical processes with different spatial and temporal scales. As ocean waves propagate from deep water to shallow water, they evolve through the interaction with the bathymetry and currents (e.g., reflection, refraction, diffraction and shoaling). Accurate descriptions of these phenomena are essential for understanding nearshore hydrodynamics and for engineering applications. For studying large-scale (multiple wavelengths) wave propagation and scattering problems, two-dimension horizontal (2DH) models are preferred over fully three-dimensional models because of the huge difference in computational efficiency. All the depth-averaged models share the same idea of reducing the computational cost greatly by removing the vertical dependence through depth-integration. The depth averaging, of course, must be based on some assumptions of the vertical flow structure and, therefore, the applications of these models are relatively restricted.

A number of 2DH depth-integrated models have been developed for nearshore applications in the past few decades. Based on the underlying assumptions and approximations, these models can be mainly classified as follows: mild-slope equation (MSE), nonlinear shallow water equation (NLSWE), Boussinesq-type of models (BTM) and Green–Naghdi (GN) models. A brief review of all these modeling approaches is given in this chapter.

3.1   Mild-slope equation models

The mild-slope equation (MSE) is a type of depth-integrated wave model that originally assumes irrotational linear waves and mild varying bottom slopes. Nevertheless, the advances in the theoretical derivation of the MSE have now extended its applications to weakly nonlinear waves, rapidly varying bottom and various types of energy dissipation (thereby called the Modified MSE or MMSE). The MSE wave model is popular to describe the wave propagation in an area where the combined effects of refraction and diffraction are important.

There are three types of MSE: the elliptic, hyperbolic and parabolic MSE. They have been developed with different aims and have continuously evolved over the past few decades. Each of the formulations presents a different range of applicability. For example, the elliptic MSE can be applied to solve for steady-state wave fields, the hyperbolic MSE can be used to solve for time-dependent wave fields, and the parabolic MSE to solve for a simplified steady-state wave field with a primary wave propagation direction (Lin, 2008). These three types are described in detail in the following sections.

3.1.1   *The elliptic MSE*

3.1.1.1   Governing equations

Taking the simple linear wave theory as a starting point, the first MSE development dates back to the 1950–1960s by Eckart (1952) and Keller (1958). The later pioneering work of Berkhoff (1972) is regarded as one of the most widely used MSE models in coastal engineering. The same MSE can be also found in Smith and Sprinks (1975), who used a different derivation approach from Berkhoff (1972).

In Berkhoff (1972) and Smith and Sprinks's (1975) works, a periodic wave train in time was considered, leading to a steady-state MSE portrayed in Eq. (3.1). This is an elliptic type of partial differential equation, which is why it is also called EMSE.

$$\nabla_{\mathbf{H}} \cdot (cc_g \nabla_{\mathbf{H}} \zeta) + k^2 cc_g \zeta = 0. \tag{3.1}$$

In this equation, $\zeta(x, y)$ is the complex-valued amplitude of the free-surface elevation $\eta(x, y, t)$; (x, y) is the horizontal position; k is the local wavenumber; c is the phase speed of the waves and c_g is the group speed of the waves.

The assumption of mild bottom slope was examined by Booij (1983), who determined that it is essentially accurate for slopes up to 1 on 3. The numerical solution of the EMSE was attempted in Berkhoff (1972) by using a finite element method. However, this method was restricted to a small domain. Various approaches have been proposed to improve the computational efficiency and to extend the EMSE to a larger domain. Examples of these are the preconditional conjugate-gradient method (Panchang *et al.*, 1991), the multigrid technique (Li and Anastasiou, 1992), the modified hybrid element method (MHEM) (Zhang, 1996), the GPBICG method (Tang *et al.*, 2004), the localized differential quadrature method (Hamidi *et al.*, 2012) and the generalized finite difference method (Zhang *et al.*, 2018).

Furthermore, by factoring out the free surface displacement or the velocity potential function in the original EMSE as a product of amplitude function and phase function can reduce the EMSE to the eikonal equation of ray theory, revealing that the correction of the effective wavenumber is due to the change of bathymetry representing both wave diffraction and wave refraction (Kaihatu, 1997; Lin, 2008).

3.1.1.2 Numerical models and applications

The original EMSE has been modified by several authors to extend its range of applicability. For example, Massel (1993) presented an extended refraction–diffraction equation for waves propagating over a bathymetry with substantial variations by including higher order terms associated with the bottom slope, the bottom curvature and the evanescent modes. Chandrasekera and Cheung (2001) provided a linear refraction-diffraction model for steep bathymetry.

Based on Kirby's (1984) EMSE, Kostense *et al.* (1989) developed a finite element model to study the effect of currents on the wave propagation in and around arbitrarily shaped harbors of variable depth. Li and Anastasiou (1992) later used a multigrid technique for water wave propagation over large areas in the presence of currents, where the combined effects of shoaling, refraction, diffraction and wave-breaking were considered. Further information on numerical modeling of wave deformation due to a current can also be found in Chen *et al.* (2005).

Yu and Togashi (1995) presented a numerical model for the transformation of narrow-banded irregular waves over an elliptic shoal based on EMSE. While using a multigrid method to solve the EMSE for irregular water wave propagation, Li *et al.* (1993) showed that their numerical scheme was very efficient and allowed spectral wave calculations to be performed over large study areas at a modest computational cost.

By combining an explicit nonlinear dispersion relation (Kirby and Dalrymple, 1986) with the weakly nonlinear MSE, Li (2001) derived a wave transformation model, accounting for the effect of nonlinear dispersion. More recently, nonlinear wave–wave interactions were further studied by Sharma *et al.* (2014).

Dalrymple *et al.* (1984) derived a modified MSE from Berkhoff (1972) in which the damping effect produced by wave-breaking was taken into consideration. Then, building on Dalrymple *et al.*'s (1984) work, Yu *et al.* (1992) developed a finite element model for wave transformation, including breaking, around a detached breakwater, in which the critical value of the breaking index follows Watanabe *et al.*'s (1984) formula. By simulating wave breaking through a modification of the wave height probability density function and by parameterizing the wave energy dissipation mechanism according to that of the hydraulic jump formulation, Lou and Massel (1994) presented a numerical model for wave propagation that combined refraction, diffraction and wave-breaking dissipation. Based on the same MSE of Lou and Massel (1994), Yu and Guan (2000) introduced a modified hybrid element numerical model with similar capabilities. Zhao *et al.* (2001) tested five wave-breaking formulations and applied them to several laboratory tests and two real-world cases. A more recent numerical model that included the wave-breaking effect can be found in Zhang *et al.* (2021).

Various applications of EMSE models include the wave model over a changing water depth within the mangrove forest (Phuoc and S., 2008), the waves propagating around a circular conical island (Hsiao *et al.*, 2010) or around axis-symmetric islands without a vertical wall along the coastline (Song *et al.*, 2015), simulation of waves in harbors (Panchang and Demirbilek, 2001; Zheng *et al.*, 2013), and simulation of waves in an area with surface protruding structures like piers and breakwaters (Zubier *et al.*, 2003).

3.1.2 The hyperbolic MSE

3.1.2.1 Governing equations

The time-dependent MSE can be derived from the EMSE (Booij, 1981; Dingemans, 1997) considering the transient evolution of the wave over a slowly varying bathymetry. The resulting equation, the transient MSE (HMSE), is a hyperbolic partial differential equation with second-order derivatives in both time and space, as shown in Eq. (3.2).

$$\frac{\partial^2 \eta}{\partial t^2} - \nabla_{\mathbf{H}} \cdot (cc_g \nabla_{\mathbf{H}} \eta) + (\sigma^2 - k^2 cc_g)\eta = 0. \tag{3.2}$$

The relationship among k, h and σ is determined by the following linear dispersion equation:

$$\sigma^2 = gk \tanh(kh). \tag{3.3}$$

The equation above presents similar characteristics to the well-known linear long-wave equation (Tuck and Hwang, 1972). Therefore, Copeland (1985) introduced the pseudo-fluxes to split the original second-order MSE into a pair of first-order hyperbolic equations, which were solved by a simple explicit finite differences scheme. A similar approach was presented in Warren *et al.* (1985). Madsen and Larsen (1987) later used a highly efficient Alternating Direction Implicit algorithm to improve the solution procedure by extracting the time-harmonic part to find the stationary solution iteratively. To improve the simulation of outgoing waves, Oliveira (2000) evaluated the effects of sponge layers on the elimination of unwanted oscillations, demonstrating an adequate performance. A compact numerical algorithm was developed to directly solve the HMSE in Lin (2004), being second-order accurate in both time and space. Building on the work by Lin (2004), a simpler and more computationally efficient numerical model was presented in Song *et al.* (2007). For more recent numerical work related to the HMSE, the reader is referred to

Bokaris and Anastasiou (2003); Chun *et al.* (2013); and Tong *et al.* (2010).

3.1.2.2 Numerical models and applications

Over many years, the range of applicability of the transient HMSE has been greatly extended. For example, Kirby (1986), Chamberlain and Porter (1995), Porter (2003), Lee *et al.* (1998), Zhang and Edge (1997) and Suh *et al.* (1997) studied the effects of different types of bathymetry (i.e., steep, rapidly varying, rippled) on wave transformation.

The nonlinearity of waves can be accounted for by either incorporating a nonlinear dispersion relationship or by introducing an additional nonlinear term (Zheng *et al.*, 2001). The nonlinear wave diffraction due to a semi-finite breakwater and breakwater gap was studied in Abohadima and Isobe (1999).

Regarding the wave–current interaction, Kirby (1984) corrected an error in Booij's (1981) derivation, and derived a model for wave–current interaction from the corrected equation, which also differs from the equation of Liu (1983). Following Kirby (1984), Ohnaka *et al.* (1989) derived a new set of time-dependent MSE to simulate wave deformation due to a current. A finite difference model was presented in Dong (1987) to deal with coastal wave propagation, including the effects of refraction, diffraction, weak reflection and wave–current interaction. Mathematical models for regular and irregular waves over a non-uniform current in a slowly varying bathymetry were derived by Hong (1996). Toledo *et al.* (2012) derived an extended HMSE model that accounted for the high-order derivatives of the bottom profile and the depth-averaged current. Wave interaction with a vertically sheared current was studied in Touboul *et al.* (2016). Additional wave–current interactions studied include Li and Zhang (1995) and Huang *et al.* (2000).

Bottom friction and the Coriolis forces were both considered to simulate tidal waves in Li *et al.* (2010), based on Kirby's (1984) time-dependent HMSE. Pan *et al.* (2001) proposed an extended HMSE, which contains the terms describing the effects of bottom friction, wind input and wave nonlinearity.

Numerical models that account for wave breaking in the surf zone were developed to predict the nearshore wave fields. Watanabe and Maruyama's (1986) numerical model combined refraction, diffraction and breaking, and showed good agreement with physical experiments of waves around a detached breakwater and a jetty. Tsai *et al.* (2001) incorporated an energy dissipation factor due to wave breaking into their model and accurately predicted wave transformation over several one-dimensional beach profiles, including uniform slope, bar and step profiles. Kubo *et al.* (1993) derived a time-dependent MSE to simulate the deformation of irregular waves due to breaking. Additional applications include the numerical model for wave transformation over a submerged permeable breakwater on a porous bottom (Tsai *et al.*, 2006), and the numerical model for vegetation-induced dissipation in a mangrove forest (Cao *et al.*, 2015).

3.1.3 *The parabolic MSE*

3.1.3.1 Governing equations

The EMSE wave models solve a boundary value problem, requiring that all boundary conditions be specified. It is, however, not always convenient to treat every wave propagation problem as a boundary value problem, especially if the boundary conditions are not well defined along the entire boundary. Additionally, for regional modeling, the required high grid resolution can be computationally prohibitive, even nowadays. To greatly reduce the computational costs, an idea is to approximate the boundary value problem to an initial value problem. This is the starting point of the parabolic version of MSE (also called PMSE). Assuming that the dominating wave propagation is the x-direction, without loss of generality, the free surface displacement can be written as

$$\zeta = A(x, y)e^{ik_{x0}x}, \tag{3.4}$$

where $A(x, y)$ is the amplitude function which takes both refraction and diffraction effects into account, and k_{x0} is the wavenumber of the far-field propagating wave in the x-direction. The parabolic approximation is then developed by making use of scaling arguments concerning the primary wave propagation direction (i.e., $\frac{\partial A}{\partial x} \ll \frac{\partial A}{\partial y}$) so as to reduce the elliptic MSE to a parabolic MSE, as follows:

$$2ik_{x0}cc_g\frac{\partial A}{\partial x} + \left[(k^2 - k_{x0}^2)cc_g + i\frac{\partial(k_{x0}cc_g)}{\partial x}\right]A + \frac{\partial}{\partial y}\left(cc_g\frac{\partial A}{\partial y}\right) = 0.$$
$$(3.5)$$

Radder (1979) was the first to derive the parabolic approximation of the MSE and applied it to the study of waves propagating over a submerged shoal. Following Radder (1979), Lozano and
Liu (1980) developed a refraction–diffraction model for linear surface water waves over a mild sloping bottom with obstacles. The
weak reflection of water waves was studied for the cases of slowly
varying water depth in Liu and Tsay (1984). The errors incurred
with the use of the parabolic approximation have been investigated
by Liu (1986). Kirby (1986a,b) developed corrections to the small-
angle parabolic approximation, allowing a greater bandwidth about
the primary wave propagation direction. Wide-angle parabolic models were developed to describe the diffraction of linear water waves
in Dalrymple and Kirby (1988), further extending the validity of the
conventional form of the parabolic wave model.

3.1.3.2   Numerical models and applications

Extensions of the PMSE focus on the wave–current interactions, wave
nonlinearity, steep or varying bottom slope, wave breaking, wave over
porous beds, effect of bottom friction and combined effects of these
factors.

For example, linear waves in the presence of large ambient currents were studied in Booij (1981) and Liu (1983), and later revis-
ited by Kirby (1984). Following Kirby (1984), Zou. *et al.* (1997)
introduced the effect of energy dissipation caused by bottom friction
and derived a practical and simple form of wave-current propagation
model over a varying bathymetry.

Using a multiple-scale perturbation method, a nonlinear Stokes
wave model over a mildly varying bathymetry was presented in Kirby
and Dalrymple (1983). Their model's accuracy was later verified by
laboratory experiments conducted by Kirby and Dalrymple (1984),
showing that including the nonlinear effects can improve the model
accuracy. Liu and Tsay (1984) also derived a refraction-diffraction

model for weakly nonlinear water waves. The inclusion of nonlinear wave effects can also be realized by the use of a corrected non-linear dispersion equation, as shown in Booij (1981). Following Booij (1981), further works (Kaihatu, 2001; Kirby, 1986a; Kirby and Dalrymple, 1986) incorporated the nonlinear dispersion relation into the parabolic model. By investigating a linear and a weakly nonlinear parabolic model for random waves Zhao and Anastasiou (1993) showed that in shallow water the inclusion of the weak nonlinearity produces much better agreement with experimental data. For additional work using PMSE to study irregular waves, the reader is referred to Lin (2001).

The Parabolic Models for water waves diffraction over an irregular bathymetry were studied in Dalrymple *et al.* (1989), in which the incident wave train was decomposed into directional modes or an angular spectrum, and the effect of the bathymetry was shown to force the generation of additional directional waves. The wave evolution over a discontinuous bathymetry was studied in Lin and Qiu (2000) through a modified PMSE model.

Bottom friction and wave breaking are two main sources of energy dissipation when waves propagate in the surf zone. Pearce and Panchang (1985) introduced a source function to replicate bottom friction into the conventional MSE of Berkhoff (1972) and resorted to the parabolic approximation. Zuo *et al.* (1993) studied wave energy loss induced by bottom friction, and showed that their model has a high accuracy for waves propagating over a long distance and a gentle slope. The effect of depth-limited wave breaking in a combined refraction/diffraction wave field was examined in Kirby and Dalrymple (1986) based on the parabolic equation. A numerical model was developed by Isobe (1987) for simulating the transformation of irregular waves due to refraction, diffraction and breaking combined. The model was based on PMSE with an additional term for energy dissipation. Numerical simulations of breaking waves based on higher-order mild slope equation were presented in Tao and Han (2001).

Regarding additional applications of the PMSE, Liu and Boissevain (1988) presented a numerical model for computing wave propagation between two breakwaters. Dalrymple *et al.* (1984) studied wave propagation in the vicinity of islands. A weakly nonlinear combined refraction and diffraction model, REF/DIF,[1] was developed by Kirby and Dalrymple (1994) and soon became a standard for modeling wave propagation. From Kirby and Dalrymple's (1994) work, Tang *et al.* (2015) proposed a numerical model for coastal wave propagation through rigid vegetation. With the intent of better approximating the physical propagation domain or boundaries, PMSE has also been transformed into a non-orthogonal coordinate system (Kirby, 1988), in the conformal coordinate systems (Kirby *et al.*, 1994), in the ray-front coordinate system (Isobe, 1986), and in the curvilinear coordinates (Cui *et al.*, 2014).

3.2   Nonlinear shallow water equation models

3.2.1   *Introduction*

The nonlinear shallow water (NLSW) equation model is usually adopted to model very long waves, e.g., tsunamis, tides and storm surges. Compared with the Boussinesq-type of model, the NLSW model is simpler because the flow is assumed to be uniform in the water column and the frequency dispersion effect is totally neglected, which has the advantage of being applicable to a very large scale because of its computational efficiency.

The nonlinear shallow-water equations were first proposed by de Saint-Venant *et al.* (1871) to model flows in open channels. The so-called Saint-Venant equations are in fact a special 2D case of the general shallow-water equations derived from depth-integrating the Navier–Stokes equations (see Chapter 5), when the horizontal length scale is much greater than the vertical length scale. Under this condition, the conservation of mass implies that the vertical velocity scale of the fluid is small compared to the horizontal velocity scale. As a result, it can be shown from the momentum equation that vertical pressure gradients are hydrostatic, and that horizontal pressure gradients are due to the displacement of the pressure surface, implying that the horizontal velocity field is constant throughout the depth of the fluid. The vertical integration process allows the vertical velocity to be removed from the equations. However, it must be noted that this does not necessarily mean that the vertical velocity is zero.

---

[1]https://www1.udel.edu/kirby/programs/refdif/refdif.html

The NLSW model has a wide range of applications. Examples include the modeling of propagation and run-up of long waves, such as tides, storm surges and tsunamis, as well as open channel and overland flows due to rainstorms or dam collapses. In atmospheric and oceanic modeling, shallow water equations are also used with an additional term, representing the Coriolis force, as a simplification of the primitive equations (see Chapter 5.1). The derivation of NLSW equations and various numerical applications can be found in Lin (2008).

3.2.2   *Governing equations*

Assuming a hydrostatic pressure distribution and constant water density, the NLSW equations can be written as

$$\frac{\partial H}{\partial t} + \frac{\partial}{\partial x}(\bar{u}H) + \frac{\partial}{\partial y}(\bar{v}H) = 0, \tag{3.6}$$

$$\frac{\partial}{\partial t}(\bar{u}H) + \frac{\partial}{\partial x}(\bar{u}^2 H) + \frac{\partial}{\partial y}(\bar{u}\bar{v}H) + D_x$$
$$= -\frac{H}{\rho}\frac{\partial p_a}{\partial x} - gH\frac{\partial \eta}{\partial x} + \frac{1}{\rho}\frac{\partial}{\partial x}\int_{-h}^{\eta}\tau_{xx}dz + \frac{1}{\rho}\frac{\partial}{\partial y}\int_{-h}^{\eta}\tau_{xy}dz$$
$$+ \frac{1}{\rho}\left[-\tau_{xx}(\eta)\frac{\partial \eta}{\partial x} - \tau_{xy}(\eta)\frac{\partial \eta}{\partial y} + \tau_{xz}(\eta)\right]$$
$$- \frac{1}{\rho}\left[\tau_{xx}(-h)\frac{\partial h}{\partial x} - \tau_{xy}(-h)\frac{\partial h}{\partial y} + \tau_{xz}(-h)\right], \tag{3.7}$$

$$\frac{\partial}{\partial t}(\bar{v}H) + \frac{\partial}{\partial x}(\bar{u}\bar{v}H) + \frac{\partial}{\partial y}(\bar{v}^2 H) + D_y$$
$$= -\frac{H}{\rho}\frac{\partial p_a}{\partial y} - gH\frac{\partial \eta}{\partial y} + \frac{1}{\rho}\frac{\partial}{\partial x}\int_{-h}^{\eta}\tau_{yx}dz + \frac{1}{\rho}\frac{\partial}{\partial y}\int_{-h}^{\eta}\tau_{yy}dz$$
$$+ \frac{1}{\rho}\left[-\tau_{yx}(\eta)\frac{\partial \eta}{\partial x} - \tau_{yy}(\eta)\frac{\partial \eta}{\partial y} + \tau_{yz}(\eta)\right]$$
$$- \frac{1}{\rho}\left[\tau_{yx}(-h)\frac{\partial h}{\partial x} - \tau_{yy}(-h)\frac{\partial h}{\partial y} + \tau_{yz}(-h)\right], \tag{3.8}$$

where $\bar{u}$ and $\bar{v}$ are the depth-averaged horizontal velocities. The fluid domain is bounded by a solid bottom ($z = -h$) and a free surface ($z = \eta$), and $H = h + \eta$ is the total water depth. Here, $\tau_{xx}, \tau_{xy}, \tau_{xz}, \tau_{yx}, \tau_{yy}, \tau_{yz}$ are various stress terms while η (or h) in the parenthesis represents the stress evaluated at the free surface (or bottom); p_a is the atmospheric pressure field. Finally, D_x and D_y are momentum dispersion terms given by

$$D_x = \frac{\partial}{\partial x} \int_{-h}^{\eta} (u - \bar{u})^2 \mathrm{d}z + \frac{\partial}{\partial y} \int_{-h}^{\eta} (u - \bar{u})(v - \bar{v}) \mathrm{d}z, \quad (3.9)$$

$$D_y = \frac{\partial}{\partial x} \int_{-h}^{\eta} (u - \bar{u})(v - \bar{v}) \mathrm{d}z + \frac{\partial}{\partial y} \int_{-h}^{\eta} (v - \bar{v})^2 \mathrm{d}z. \quad (3.10)$$

By assuming the horizontal velocity to be depth-uniform in the water column, i.e., $u = \bar{u}$, $v = \bar{v}$ and τ_{xx} and τ_{xy} to be constant in the vertical direction, the momentum equations can be simplified into the following form:

$$\frac{\partial}{\partial t}(\bar{u}H) + \frac{\partial}{\partial x}(\bar{u}^2 H) + \frac{\partial}{\partial y}(\bar{u}\bar{v}H)$$

$$= -\frac{H}{\rho}\frac{\partial p_a}{\partial x} - gH\frac{\partial \eta}{\partial x} + \frac{H}{\rho}\frac{\partial \tau_{xx}}{\partial x} + \frac{H}{\rho}\frac{\partial \tau_{xy}}{\partial y}$$

$$+ \frac{1}{\rho}\tau_{xz}(\eta) - \frac{1}{\rho}\tau_{xz}(-h), \quad (3.11)$$

$$\frac{\partial}{\partial t}(\bar{v}H) + \frac{\partial}{\partial x}(\bar{u}\bar{v}H) + \frac{\partial}{\partial y}(\bar{v}^2 H)$$

$$= -\frac{H}{\rho}\frac{\partial p_a}{\partial y} - gH\frac{\partial \eta}{\partial y} + \frac{H}{\rho}\frac{\partial \tau_{yx}}{\partial x} + \frac{H}{\rho}\frac{\partial \tau_{yy}}{\partial y}$$

$$+ \frac{1}{\rho}\tau_{yz}(\eta) - \frac{1}{\rho}\tau_{yz}(-h). \quad (3.12)$$

where $\tau_{xz}(\eta)$ and $\tau_{xz}(-h)$ are surface and bottom shear stresses, respectively, which usually come into the equations as source terms to model surface wind stress and bottom friction effects. The above equations have been used by Dean and Dalrymple (1991).

Another simplified version of NLSW equations can be obtained if we further assume the viscous effects, surface and bottom forcing terms are negligible, i.e.

$$\frac{\partial}{\partial t}(\bar{u}H) + \frac{\partial}{\partial x}(\bar{u}^2 H) + \frac{\partial}{\partial y}(\bar{u}\bar{v}H) = -gH\frac{\partial\eta}{\partial x}, \qquad (3.13)$$

$$\frac{\partial}{\partial t}(\bar{v}H) + \frac{\partial}{\partial x}(\bar{u}\bar{v}H) + \frac{\partial}{\partial y}(\bar{v}^2 H) = -gH\frac{\partial\eta}{\partial y}. \qquad (3.14)$$

In the above equations, the convection terms are written in the conservative form. Making use of the continuity equation, Eq. (3.6), the equivalent NLSW equations in non-conservative form can be obtained as follows:

$$\frac{\partial\bar{u}}{\partial t} + \bar{u}\frac{\partial\bar{u}}{\partial x} + \bar{v}\frac{\partial\bar{u}}{\partial y} = -g\frac{\partial\eta}{\partial x}, \qquad (3.15)$$

$$\frac{\partial\bar{v}}{\partial t} + \bar{u}\frac{\partial\bar{v}}{\partial x} + \bar{v}\frac{\partial\bar{v}}{\partial y} = -g\frac{\partial\eta}{\partial y}. \qquad (3.16)$$

Analytical solutions to the NLSW equations are limited to a few simple cases, and for most practical applications, numerical methods need to be used. Among the numerous numerical work, Godunov-based finite-volume methods are widely used for solving the NLSW equations that have a hyperbolic nature, as these methods are proven to be robust and accurate in capturing discontinuities such as broken bores. Godunov scheme is to solve exact or approximate Riemann problems at each inter-cell boundary. Enormous attention has thus been devoted to determining a proper Riemann solver. A comprehensive review on the Riemann solver for NLSW equations can be found in Wu and Cheung (2008). For example, Roe (1981) proposed an approximate solver for Euler's equation through the solution of a linearized Riemann problem. The Roe-type Riemann solver was then applied to the NLSW equations by Glaister (1988) and was extended for wave run-up, overtopping and regeneration problems (Dodd, 1998; Hubbard and Dodd, 2002). Using a Roe-type linearization in a quasi-steady wave-propagation algorithm enables the balance of the source terms and flux gradients in high resolution Godunov methods (LeVeque, 1998). Alternatively, Harten *et al.* (1983) suggested an approach for the solution of the Riemann problem through an approximation to the numerical flux by three constant states separated by two waves with constant speeds, which can be calculated by Einfeldt's (1988) algorithms. This approach was also known as the HLL or HLLE Riemann solver. Using an HLL Riemann solver, Hu *et al.* (2000) proposed a one-dimensional shallow-water model for wave propagation, run-up and overtopping, and Zhou *et al.* (2001) described a well-balanced model with the surface gradient method. In addition to the approximated Riemann solvers, the exact solver was adopted by some researchers through numerical iterative schemes (Brocchini *et al.*, 2001; Pan *et al.*, 2007; Wei *et al.*, 2006). Since the exact Riemann solver needs to be applied at all the cell interfaces every time step with an iterative scheme, this restricts its implementation as the approximate solvers. The details of Riemann solvers and numerical methods can be found in the book of Toro (2013). More recent works on the numerical methods for NLSW equations can be found in Xing (2017), Kurganov (2018) and Castro and Semplice (2019).

3.2.3   *Numerical models and applications*

Because of the great simplification of the vertical profile of the horizontal velocity, the dispersive effects are totally ignored in NLSWE. Thus, NLSWE can be only applied to model very long waves or currents). Some major applications of the NLSWE are summarized as follows:

3.2.3.1   Tides

Tides are extremely long waves that have a period up to hours. One of the main applications of NLSW models is for the simulation of tidal currents in coastal waters, e.g., Westerink *et al.* (1992). The NLSW model can be applied to simulate tidal flows by specifying the lateral boundaries in the open sea away from the coastline. Based on the available tidal tables and their interpolated values on the boundaries, the time histories of tidal heights can be specified at all lateral boundaries. The tidal flows inside the computational domain are driven by the difference of tidal (free surface elevation) heights at the different boundaries, e.g., Shankar *et al.* (1997). The numerical modeling of tides using NLSW models is quite accurate since the driving force is definite (astronomical effects) and the tides are of a strict shallow-water waves-type because of their extremely long periods.

3.2.3.2 Tsunamis

Tsunami modeling is another important field of application of NLSW models. Tsunamis are often generated by geophysical processes such as an underwater earthquake or landslide, which may cause severe coastal flooding. The NLSW models are widely used for modeling tsunami waves because of their very long wavelengths. To model a tsunami, three main processes need to be simulated, namely tsunami generation due to earthquakes or landslides, tsunami propagation in deep ocean, and tsunami run-up and inundation along coastlines.

A number of tsunami models have been developed over the past few decades based on the NLSW equations. For example, the Method Of Splitting Tsunami (MOST) model, developed by Titov of the Pacific Marine Environmental Laboratory and Synolakis of the University of Southern California (Titov and Synolakis, 1998), was adopted by the National Oceanic and Atmospheric Administration (NOAA). MOST is a suite of numerical simulation codes capable of simulating three processes of tsunami evolution: earthquake, transoceanic propagation and inundation of dry land, which has been extensively tested against a number of laboratory experiments and was successfully used for simulations of many historical tsunamis. Besides, TUNAMI-N2[2] (Tohoku University's Numerical Analysis Model for Investigation of Near-field tsunamis) model (Shuto *et al.*, 1990), developed by the Disaster Control Research Center of Tohoku University in Japan, is another well-known tsunami model. It uses a compact leapfrog FD scheme in time and space discretization, which is efficient and accurate for modeling linear long waves. This model has been widely used in modeling tsunami propagation and run-up. Another example is the tsunami model COMCOT (COrnell Multigrid-COupled Tsunami) developed at Cornell University, which has been used to simulate solitary wave run-up on a circular island (Liu *et al.*, 1995) and real tsunami run-up events (Liu, 1995).

3.2.3.3 Storm surge and ocean circulation

The water surface can be significantly distorted under the influence of wind blowing. Surface set-up (set-down) can be observed if the

---

[2]https://www.tsunami.irides.tohoku.ac.jp/media/files/_u/project/manual-ver-3_1.pdf

wind blows on-shore (offshore). If the wind direction is parallel to the shoreline, longshore currents can be generated. Storm surge is the abnormal rise in seawater level during a storm, measured as the height of the water above the normal predicted astronomical tide. The surge is caused primarily by a storm's winds pushing water onshore. The amplitude of the storm surge at any given location depends on the orientation of the coast line with the storm track; the intensity, size and speed of the storm; and the local bathymetry. Along the coast, storm surge is often the greatest threat to life and property from a hurricane. Hurricane Katrina (2005) is a prime example of the damage and devastation that can be caused by storm surge.

Some well-developed storm surge models include the SLOSH model[3] (the Sea, Lake and Overland Surges from Hurricanes); the ADvanced CIRCulation (ADCIRC)[4] coastal circulation and storm surge model; and (CH3D)-SMSS,[5] the Storm Surge Modeling System with Curvilinear-grid Hydrodynamics in 3D and the Finite-Volume, Primitive Equation Community Ocean Model (FVCOM).[6] Additional storm surge models include Delft3D,[7] the Japan Meteorological Agency (JMA) storm surge model and SCHISM[8] (Semi-implicit Cross-scale Hydroscience Integrated System Model). Some of these models are briefly introduced as follows.

SLOSH is a model (Jelesnianski, 1992) that was developed by the National Weather Service (NWS) in the 1990s and is still the operational model used by the National Hurricane Center (NHC) due to its computational efficiency. The SLOSH model represents the domain with a mesh (see website[9]), which can resolve ocean bathymetry and land features in great detail, but does not extend to the open ocean.

The ADCIRC model (Luettich *et al.*, 1992) was developed at the University of North Carolina. It is a more complex and capable, thus more time-consuming, storm surge model than SLOSH. ADCIRC makes use of a highly flexible mesh system to produce

---

[3]https://www.nhc.noaa.gov/surge/slosh.php
[4]https://adcirc.org/
[5]https://aces.coastal.ufl.edu/CH3D/
[6]http://fvcom.smast.umassd.edu/fvcom/
[7]https://oss.deltares.nl/web/delft3d/
[8]http://ccrm.vims.edu/schismweb/
[9]http://www.hurricanescience.org/science/forecast/researchmodels/

storm surge simulations. The model can better simulate tides propagated from the open ocean and is also capable of resolving very detailed bathymetries in the coastal region. After Hurricane Katrina, the Interagency Performance Evaluation Taskforce (IPET) used the primarily two-dimensional ADCIRC model to simulate Katrina's storm surge and produce coastal flood maps for a storm with a 1% and 0.2% annual chance of occurrence in a given year for the New Orleans region. ADCIRC is also being used by the U.S. Federal Emergency Management Association (FEMA) to produce Flood Insurance Rate Maps (FIRMs) in several coastal regions. A storm surge and wave guidance system is being developed based on the ADCIRC model that can be used for coastal emergency risk assessment.[10]

SCHISM (Semi-implicit Cross-scale Hydroscience Integrated System Model) is an open-source community-supported modeling system based on unstructured grids, designed for seamless simulation of 3D baroclinic circulation across creek-lake-river-estuary-shelf-ocean scales (Zhang and Baptista, 2008; Zhang *et al.*, 2016b). It uses a highly efficient and accurate semi-implicit finite-element/finite-volume method with Eulerian–Lagrangian algorithm to solve the Navier–Stokes equations (in hydrostatic form), in order to address a wide range of physical and biological processes. The numerical algorithm judiciously mixes higher-order with lower-order methods, to obtain stable and accurate results in an efficient way. Mass conservation is enforced with the finite-volume transport algorithm. It also naturally incorporates wetting and drying of tidal flats.

The Delft3D modeling suite (Lesser *et al.*, 2004) is currently used by the Naval Oceanographic Office for regional and nearshore applications and more recently to forecast surge and inundation (Veeramony *et al.*, 2014) in their Coastal Surge and Inundation Prediction System (CSIPS). The hydrodynamic solver discretizes the shallow water equations based on finite differences, and additional modules, such as the transport solver, making use of finite volume discretizations. Two- (2D) and three-dimensional (3D) space is meshed using rectangular, curvilinear or spherical grids. Included features comprise "wind shear, wave forces, tidal forces, density-driven flows and stratification due to salinity and/or temperature gradients, atmospheric pressure changes, drying and flooding of intertidal flats"

---

[10]https://cera.coastalrisk.live/

(Lesser *et al.*, 2004), making Delft3D suitable to model coastal, river and estuarine areas. The simulations can be set up using the open-source tool Delft Dashboard (Van Ormondt *et al.*, 2020). Examples of applications include modeling surge and inundation during Hurricane Ike, which impacted the northern Gulf of Mexico in September 2008 (Veeramony *et al.*, 2017).

3.2.3.4   Open channel flows and river flows

The NLSW models can also be applied to study open channel and river flows. Open-channel flow can be classified into steady and unsteady flows depending on whether the flow field changes with time. Unsteady open channel flows share a lot of similarities with long waves, with varying free surface elevation and near hydrostatic pressure. However, for open channel flows, gravity is the driving force, rather than the restoring force, as happens for waves. The ratio of the inertial to gravitational forces acting on the flow is represented by the dimensionless Froude Number (Fr). Depending on whether the Froude number is greater or less than unity, the flow is classified as supercritical or subcritical respectively. The hydraulic behavior of open-channel flow varies significantly depending on whether the flow is critical, subcritical, or supercritical. Numerous works can be found in numerical modeling of open channel flow using NLSW equations model. The theory and numerical computation of open channel flow can be found in Chaudhry (2007). On the other hand, both commercial software and freeware are available for the modeling of river hydraulics using the NLSW approach. For example, MIKE HYDRO River developed by DHI is a commercial software that can be used to carry out flood analyses, investigate flood alleviation options and tackle hydraulic design within river networks, including canal systems and operational structures.

For additional applications of NLSW, the reader is referred to Delestre *et al.* (2013) and Kirstetter *et al.* (2021).

3.3   Boussinesq-type of models

3.3.1   *Introduction*

Another approach to model the propagation of waves over a large area is the Boussinesq-type of models (BTMs). These models are derived from the pioneering work by Boussinesq (1872) the late 1800s in the context of relatively long waves.

There are two basic assumptions adopted in the classic BTMs: the weak nonlinearity and the weak dispersion. For example, the classic BTMs (Peregrine, 1967) require that frequency dispersion and nonlinearity parameters are not only small, but also in the same order of magnitude, i.e., $\mathcal{O}(\lambda^2) = \mathcal{O}(\epsilon) \ll 1$, where λ is the ratio between water depth and characteristic wavelength (h/L) and ϵ is the ratio between wave amplitude and water depth (A/h). Moreover, the range of applicability of the classic BTMs is limited to relatively shallow water (i.e., $\lambda < 0.175$ or $kh < 1.1$, with $k = 2\pi/L$ being the wavenumber). Given these, the accuracy of the classic BTMs is $\mathcal{O}(\lambda^2)$, as shown in what follows:

$$\frac{\partial \eta}{\partial t} + \nabla_{\mathbf{H}} \cdot \left[(h + \eta)\bar{\mathbf{U}} \right] = \mathcal{O}(\epsilon^2, \epsilon\lambda^2, \lambda^4), \qquad (3.17)$$

$$\frac{\partial \bar{\mathbf{U}}}{\partial t} + (\bar{\mathbf{U}} \cdot \nabla_{\mathbf{H}})\bar{\mathbf{U}} + g\nabla_{\mathbf{H}}\eta - \frac{1}{2}h\frac{\partial}{\partial t}\nabla_{\mathbf{H}}$$
$$\times \left[\nabla_{\mathbf{H}} \cdot (h\bar{\mathbf{U}}) \right] + \frac{1}{6}h^2\frac{\partial}{\partial t}\nabla_{\mathbf{H}}(\nabla_{\mathbf{H}} \cdot \bar{\mathbf{U}})$$
$$= \mathcal{O}(\epsilon^2, \epsilon\lambda^2, \lambda^4), \qquad (3.18)$$

where $\bar{\mathbf{U}}$ is the depth-averaged velocity vector, i.e., $\bar{\mathbf{U}} = (\bar{u}, \bar{v})$. Since the development of the classic BTMs, several efforts have been made to extend the classic BTMs capabilities to waves in intermediate and deep waters by improving the linear frequency dispersion accuracy, while the nonlinearity is still kept small (e.g., Madsen *et al.*, 1992; Nwogu, 1993; Schäffer and Madsen, 1995). This has lead to the so-called extended BTMs. To consider the nonlinear effects, a series of fully nonlinear and weakly-dispersive wave models have been developed by treating the nonlinearity parameter ϵ as $\mathcal{O}(1)$. Wei *et al.* (1995), starting from Euler's equations, developed a model, which is still of $\mathcal{O}(\lambda^2)$ accuracy. Following Wei *et al.* (1995), Gobbi *et al.* (2000) included the $\mathcal{O}(\lambda^4)$ terms in their model equations. Consequently, the linear wave frequency dispersion of Gobbi *et al.*'s (2000) model is accurate to $kh \approx 6$. However, the partial differential equations in the higher-order extended BTMs contain higher spatial derivatives (> 3), which requires additional boundary conditions and also higher computational effort. Additional works with a similar level of accuracy (Agnon *et al.*, 1999; Madsen and Agnon, 2003) also encounter numerical difficulties due to the higher order spatial derivatives. These numerical difficulties limit the further extension of the extended BTMs.

To avoid solving the complexity involved in the higher spatial derivatives, Lynett and Liu (2004a,b) offered an alternative approach, in which they incorporated the multi-layer formulation with the fully nonlinear and weakly dispersive wave approximations. In their model, the water column is first divided into several layers with both the horizontal velocity and the pressure being matched at the interfaces between layers. A quadratic horizontal velocity profile and linear vertical velocity profile are employed by assuming weak $(O(\lambda^4))$ horizontal vorticity in each layer. More accurate velocity profiles in the water column and better representation of linear frequency dispersion relation can be obtained by increasing the number of layers. For example, the two-layer model exhibits accurate linear and nonlinear characteristics up to $kh \approx 6.5$. It is important to note that the highest spatial derivative in the multi-layer models is always third. Lynett (2006) later included an eddy-viscosity type of breaking model into Lynett and Liu (2004a), and showed significant benefits close to the breaker line. These works laid the foundations of the free surface wave model COULWAVE,[11] of which the primary applications include landslide tsunami generation and propagation, nearshore tsunami evolution and inundation, and nearshore wind wave modeling.

A more recent work for the multi-layer wave model can be found in Liu *et al.* (2018), who constructed a different type of multi-layer wave model for simulating highly dispersive and highly nonlinear waves over a slowly varying bottom. The model was developed specifically for potential flow and the governing equations were written in terms of the variables on the free surface (Agnon *et al.*, 1999). Following Lynett and Liu (2004a), Liu *et al.* (2018) also divided the water column into multiple layers. However, the long-wave assumption was relaxed, and both horizontal velocity and vertical velocity were formulated using Taylor series expansions at a certain elevation.

---

[11] http://isec.nacse.org/models/coulwave_description.php

Thus, the vertical velocity must be solved explicitly instead of being expressed in terms of derivatives of horizontal velocity. Following Madsen *et al.* (2002, 2003), the linear and nonlinear properties were further enhanced and the resulting model demonstrated high accuracy in extremely deep waters. For example, their two-layer model is accurate up to $kh \approx 53$ with 1% error in terms of linear frequency dispersion. However, we reiterate here that this model is restricted to potential flow and the water depth can only be slowly varying. Moreover, the multi-layer formulation inevitably requires solving more equations. As an example, if a one-dimensional horizontal (1DH) case with N layers is solved, $(N+1)$ and $(2N+5)$ equations need to be solved for the models of Lynett and Liu (2004a) and Liu *et al.* (2018), respectively.

Another line of the development of the extended BTMs can be traced to Chen and Liu (1995), who rederived Nwogu's (1993) BTMs in terms of a velocity potential (irrotational) on an arbitrary elevation and accounting for the free surface displacement. Considering the effect of the leading order vertical vorticity ($O(1)$), Hsiao *et al.* (2002) produced an extended BTMs for fully nonlinear water waves ($\epsilon = \mathcal{O}(1)$) propagating over a permeable bed. Chen (2006) later introduced a set of extended BTMs that accounted for the second-order effects of vertical vorticity, such that their model is suitable for water waves and wave-induced currents with strong vertical vorticity. A high-order adaptive time-stepping Total Variation Diminishing (TVD) solver for the fully nonlinear BTMs of Chen (2006) was presented in Shi *et al.* (2012), extended to include a moving reference level as in Kennedy *et al.* (2001). Consequently, the FUNWAVE[12] model, initially developed by Kirby *et al.* (1998), was updated to the FUNWAVE-TVD[13] (the TVD version of FUNWAVE), with an accuracy of ($O(\lambda^2)$). This wave model can be used for modeling surfzone-scale optical properties (see Shi *et al.*, 2010), the modeling of tsunami waves in both a global/coastal scales for predicting coastal inundation and a basin scale for wave propagation. Furthermore, the recent developments of FUNWAVE-TVD include

---

[12]http://www1.udel.edu/kirby/programs/funwave/funwave.html
[13]https://fengyanshi.github.io/build/html/index.html

ship-wave generation (Mattosinho *et al.*, 2021), meteo-tsunami generation (Woodruff *et al.*, 2018) and sediment transport and morphological changes (Malej *et al.*, 2019; Tehranirad *et al.*, 2017).

3.3.2 *Numerical models and applications*

Practical applications of BTMs usually involve complex physics (i.e., wave breaking, bottom friction, subgrid-scale lateral mixing effects) that need to be incorporated via *ad hoc* models. For example, the accurate description of wave run-up and wave motion over natural coasts is necessary to simulate the land–sea interface realistically. An early method that was widely used was based on the so-called "slot method" (Tao, 1983a,b) in which the beach contains narrow "slots" so that the water level can be below the beach elevation (Chen *et al.*, 2000; Kennedy *et al.*, 2000; Madsen *et al.*, 1997). However, this approach was noisy and interfered with shoaling and breaking. An alternate approach was proposed by Lynett and Liu (2002), in which the shoreline was identified by a linear landward extrapolation from water elevations at the shallowest "wet" grid points. Although this approach was stable, it led to mass conservation problems. Through shifting to the use of finite volume schemes that came to dominate the Boussinesq world in the early 2000s, the Godunov-style schemes used an approximate solution of the Riemann problem to reconstruct velocities at cell boundaries, and the moving shoreline boundary condition was then implemented using the wetting–drying algorithm with the adjusted wave speed of the Riemann solver in a straightforward manner (Shi *et al.*, 2012).

Various attempts have been made to introduce wave breaking effects into BTMs. Early treatments include eddy viscosity models (Kennedy *et al.*, 2000; Zelt, 1991), "roller" breaking model (Schäffer *et al.*, 1993) or detailed vorticity models (Veeramony and Svendsen, 2000). Recent progress in this area includes the use of a breaking criterion (Froude number) in either eddy viscosity or roller models (Okamoto and Basco, 2006) or the use of a breaking celerity index as shown in D'Alessandro and Tomasicchio (2008). Likewise, Tonelli and Petti (2009) used a hybrid approach to shift from BTMs to nonlinear shallow water (NLSW) equations in which dispersive effects are turned off when a Froude number criterion, based on the ratio of crest-to-trough wave height to water depth in front of the crest, is reached. A similar approach but different criterion based on the gradient of momentum was used in Roeber and Cheung (2012).

In addition to the energy dissipation due to wave breaking, which is assumed to be strongly localized on the front face of the breaking wave, the horizontally distributed eddy viscosity resulting from subgrid turbulent processes may become an important factor that influences the flow pattern of the wave-generated current field. To account for the effect of resultant eddy viscosity on the underlying flow, Chen *et al.* (2000, 1999) incorporated a Smagorinsky-type subgrid model (Smagorinsky *et al.*, 1965) into the BTMs.

The bottom friction is often treated using standard quadratic formulations based on depth-averaged velocities with calibrated friction coefficients. To address the presence of turbulence and vorticity in a more direct and physically consistent manner, Kim *et al.* (2009) developed a more comprehensive treatment of the depth-averaged flow field by taking into account the presence of turbulence and vorticity. From this study, extensive work were achieved and can be found in Kim and Lynett (2011) and Kim (2015).

Furthermore, to apply the BTMs in complex geometries, the classic approach of employing stair-step boundaries or using rectangular grids in existing Cartesian grid finite difference codes is somewhat problematic. Alternatively, Shi *et al.* (2001) rederived the BTMs in generalized coordinates so as to adapt computations to irregularly shaped shorelines, such as harbors, bays and tidal inlets, and made computations more efficient in large near-shore regions. The wind effects were incorporated into BTMs by Chen *et al.* (2004) through parameterizing the wind stress as a function of wave steepness and wind speed.

Wider applications of BTMs include solving rip currents (Chen *et al.*, 1999; Wang *et al.*, 2018a), current–wave interactions (Yang and Liu, 2020), longshore currents (Choi *et al.*, 2015) and tsunami-induced currents (Lynett *et al.*, 2017). Moreover, Dong *et al.* (2010) and Gao *et al.* (2018) studied wave-induced harbor resonance; Grilli *et al.* (2020) assessed the impact of extreme storms and derived coastal risk indicators; Ning *et al.* (2019) studied wave propagation and run-up over fringing reefs. Numerous works for the natural hazards assessment using BTMs can be found in [Watts *et al.* (2003)], [Day *et al.* (2005)], [Ioualalen *et al.* (2007)], [Tappin *et al.* (2008)], [Waythomas *et al.* (2009)], [Abadie *et al.* (2012)], Ha *et al.* (2014), [Shelby *et al.* (2016)], [Nemati *et al.* (2019)], and [Paris and Ulvrova (2019)]. Finally, for further BTMs applications to coastal processes across a wide range of scales, the reader is referred to Kirby (2016).

3.4   Green–Naghdi models

3.4.1   *Introduction*

The Green–Naghdi (GN) equations modeling is another approach for the propagation of fully nonlinear and weakly dispersive waves. Green *et al.* (1974) and Green and Naghdi (1976a,b) first derived the GN equations based on continuum mechanics theory and the concept of "directed fluid sheets," which has its origin in the nonlinear plate and shell theory. Ertekin *et al.* (1986) rederived the GN equations by assuming a uniform horizontal velocity and a linear vertical velocity profile in the water column, without using any small perturbation parameters and the irrotational flow assumption. This fully nonlinear GN model has been successfully applied to simulate ship waves generated by a prescribed moving free surface pressure gradient. The GN equations (Ertekin *et al.*, 1986) can be shown to be the same as the model equations derived by Serre (1953), Su and Gardner (1969) and Barthélemy (2004). Moreover, Lannes and Bonneton (2009) verified that the GN equations (Ertekin *et al.*, 1986; Green and Naghdi, 1976a) are equivalent to the fully nonlinear and weakly dispersive Boussinesq model (Wei *et al.*, 1995).

Another development of GN equations is due to Shields and Webster (1988), who used the Kantorovich method (Kantorovich and Krylov, 1958) to reduce the dimensionality of the problem. Shields and Webster (1988) derived the GN equations for steady two-dimensional (horizontal and vertical) flows by assuming the horizontal velocity profile as a second degree polynomial of the vertical coordinate. Demirbilek and Webster (1992, 1999) derived the so-called Level I and Level II (Level K corresponds to a $(K-1)$ degree horizontal velocity assumption in the referenced work) GN equations and further applied these equations to simple shallow water and nonlinear wave propagation problems. Extending the work of Demirbilek and Webster (1999), Webster *et al.* (2011) provided an improved derivation of the GN equations for higher polynomial approximations in the velocity profiles.

In developing the above GN models, the velocity components are first approximated as a finite series of inner products of two sets of functions: a pre-selected shape or trial functions, depending only on the vertical coordinate, and unknown functions to be determined, depending on horizontal coordinates and time. Since these solutions are not exact, residuals can be calculated from the governing equations. The Galerkin method is then applied. Thus, the same set of shape functions is employed as the weighting functions in evaluating the residuals.

Most recently, Yang and Liu (2020) derived two sets of depth-integrated wave-current models based on the Euler equations written in terms of a σ-coordinate (see Fig. 5.1), mapping the water column into a constant interval in σ. The total horizontal velocity in the water column was assumed to be a polynomial, which is a truncated infinite series of products of prescribed shape functions of σ and unknown functions of the horizontal coordinates and time. As a result, Galerkin models (GK) and subdomain models (SK) of different degree of approximations were derived by minimizing the residuals of the horizontal momentum equations with respect to either the Galerkin method or the subdomain method, respectively. A Stokes wave-type analysis in terms of various linear and nonlinear wave properties was conducted to demonstrate the skills of the model with different approximations. For example, the Galerkin models and subdomain models with a fourth-degree polynomial approximation for the horizontal velocity profile were found applicable at very deep waters (up to $kd \approx 16.9$ and 25.1) with less than 2% error in linear wave phase speed. It has been independently shown in Yang and Liu (2020) that the Galerkin models share the same accuracy with the GN model developed by Webster *et al.* (2011). However, for the same degree of polynomial approximation on the horizontal velocity, subdomain models show superior performance in terms of all of the wave properties, compared with the Galerkin models. Both models are capable of reproducing the velocity profiles of linear wave theory with slightly different deviations. Both mathematical models were implemented numerically to study various wave transformation benchmark experiments and newly conducted laboratory experiments of a self-focusing wave group in both 1DH and 2DH. The models of different approximations were tested and their applicable ranges were identified. Excellent agreements were found between the converged numerical results and experimental data, not only for free surface elevations but also for horizontal and vertical velocities.

A further development of the models derived in Yang and Liu (2020) is to include the effects of vertically arbitrarily sheared currents (Yang and Liu, 2022), and the resulting model (C-SK) is an extension of the SK model. Since the total horizontal velocity (the combination of current and wave velocities) was approximated in Yang and Liu (2020), a higher degree polynomial must be used to simulate deeper water waves and/or currents with complex profiles in the water column. To overcome this shortcoming, a new approach has been proposed so that the vertical profile of the current can be adopted in the model without further approximation. The new approach decomposes the solutions for the horizontal velocity into two unknown components: the first component adopts the vertical profile of the prescribed steady-state current in the σ-coordinate, which contains the unknown free surface elevation. The constraint on the first solution component is that, in the absence of waves, it reduces to the prescribed steady-state current field. The second component of the horizontal velocity is approximated in the polynomial form, in a similar way as shown in Yang and Liu (2020). Euler equations and boundary conditions are used to constrain the total solutions. A theoretical analysis is conducted to study the frequency dispersion relation of linear waves on currents with an exponential vertical profile and the results are compared with numerical solutions of the Rayleigh equation. Using the new models, validations and investigations are then conducted for periodic waves and solitary waves on currents with an arbitrary profile in one-dimensional horizontal (1DH) space. Furthermore, the new models are applied to wave–current interactions in two-dimensional horizontal (2DH) space. Two scenarios are considered: (1) wave propagation over a vortex-ring-like current and (2) obliquely incident wave propagation over a 3D sheared current on a varying bathymetry. The vertical and horizontal shear of the current have significant effects on modifying various wave properties, which are well captured by the C-SK model. However, the time-averaged velocity under wave–current interaction shows small differences with the prescribed current velocity, except in the region between the wave trough and crest.

3.4.2 *Governing equations*

The governing equations for the simplest one-dimensional Green–Naghdi equations (Su and Gardner, 1969) on a flat bottom are

$$\frac{\partial H}{\partial t} + \frac{\partial (H\bar{u})}{\partial x} = 0, \tag{3.19}$$

$$\frac{\partial (H\bar{u})}{\partial t} + \frac{\partial}{\partial x}\left[H\bar{u}^2 + \frac{gH^2}{2} - \frac{H^3}{3}\left(\frac{\partial^2 \bar{u}}{\partial x \partial t}\right.\right.$$

$$\left.\left. + \bar{u}\frac{\partial^2 \bar{u}}{\partial x^2} - \left(\frac{\partial \bar{u}}{\partial x}\right)^2\right)\right] = 0, \tag{3.20}$$

where H is the total water depth and $\bar{u}$ is the horizontal velocity, which is assumed to be uniform over the water depth. For models employing higher-degree polynomial approximation on the horizontal velocity profile, the model equations become increasingly complex. As an illustration, the linear terms in the governing equations on a constant water depth (h) for the *S2* model developed in Yang and Liu (2020), which assumes the horizontal velocity to be linear in the σ-coordinate, i.e., $u = u_0 + u_1\sigma$, are shown here for brevity,

$$\frac{\partial \eta}{\partial t} + h\frac{\partial u_0}{\partial x} + \frac{1}{2}h\frac{\partial u_1}{\partial x} + NL = 0, \tag{3.21}$$

$$\frac{\partial u_0}{\partial t} + \alpha_1 \frac{\partial u_1}{\partial t} + g\frac{\partial \eta}{\partial x} + \beta_1 h^2 \frac{\partial^3 u_0}{\partial^2 x \partial t}$$

$$+ \gamma_1 h^2 \frac{\partial^3 u_1}{\partial^2 x \partial t} + NL = 0, \tag{3.22}$$

$$\frac{\partial u_0}{\partial t} + \alpha_2 \frac{\partial u_1}{\partial t} + g \frac{\partial \eta}{\partial x} + \beta_2 h^2 \frac{\partial^3 u_0}{\partial^2 x \partial t}$$

$$+ \gamma_2 h^2 \frac{\partial^3 u_1}{\partial^2 x \partial t} + NL = 0, \tag{3.23}$$

where

$$\alpha_1 = \frac{1}{2}c_1, \quad \beta_1 = -\frac{1}{2} + \frac{1}{6}c_1^2, \quad \gamma_1 = -\frac{1}{6} + \frac{1}{24}c_1^3,$$

$$\alpha_2 = \frac{1}{2}(1 + c_1), \quad \beta_2 = -\frac{1}{3} + \frac{1}{6}c_1 + \frac{1}{6}c_1^2,$$

$$\gamma_2 = -\frac{1}{8} + \frac{1}{24}c_1 + \frac{1}{24}c_1^2 + \frac{1}{24}c_1^3, \tag{3.24}$$

and NL denotes all the different nonlinear terms in the full expressions. The unknowns in Eqs. (3.21), (3.22) and (3.23) are u_0, u_1 and η. It has been shown that the model developed in Yang and Liu (2020) based on the subdomain method demonstrates better model performance compared with the Green–Naghdi equations (Webster *et al.*, 2011) of the same polynomial approximation on the horizontal velocity profile by optimizing the free parameter, e.g., c_1 in the above equations, with known analytical wave properties. Lastly, it should be noted that the highest spatial derivatives of these types of equations is always third regardless of the approximations on the horizontal velocity profile, which is one of the advantages of the present models compared with higher-order BTMs.

3.4.3   *Numerical models and applications*

Demirbilek and Webster (1992, 1999) were the first works that applied Green–Naghdi equations for engineering applications. Because of the great complexity of the governing equations, they only derived model equations up to a quadratic polynomial assumption on the horizontal velocity profiles for both steady and unsteady waves. The resulting numerical model has been successfully applied to simple wave-structure interaction and wave shoaling problems. Zhao *et al.* (2014) numerically implemented the Green–Naghdi equations derived in Webster *et al.* (2011) in a one-dimensional horizontal

(1DH) domain, which can be extended to high degree polynomial approximations on the horizontal velocity profiles. The resulting numerical model was applied to various coastal wave transformation problems, especially nonlinear dispersive wave transformation over submerged bars. An extension to three-dimensional applications was achieved in Zhao *et al.* (2015), who studied several wave transformation problems over uneven bottoms. Zhang *et al.* (2014b, 2016a) reported the surf-zone applications of Green–Naghdi equations, in which eddy viscosity is used to describe both wave breaking dissipation and bottom friction, with breaking viscosities derived from the turbulent kinetic energy equation coupled with the Green–Naghdi model derived in Zhang *et al.* (2013). Furthermore, waves interacting with depth-uniform currents are studied in Duan *et al.* (2016), and the results are compared with BTMs. Duan *et al.* (2018) also obtained the steady solution of a solitary wave propagating in the presence of a linear shear background current numerically.

On the other hand, several numerical investigations have been conducted on the Green–Naghdi equations. A hybrid numerical method using a Godunov-type scheme was proposed to solve the Green–Naghdi model describing dispersive waves by Le Métayer *et al.* (2010). Bonneton *et al.* (2011) proposed a hybrid finite volume and finite difference splitting approach to study fully nonlinear and weakly dispersive Green–Naghdi modeling for shallow water waves of large amplitude. Chazel *et al.* (2011) studied a hybrid finite-volume and finite-difference splitting approach to investigate the ability of a Green–Naghdi model to reproduce strongly nonlinear and dispersive wave propagation. Panda *et al.* (2014) introduced a local discontinuous Galerkin method for Boussinesq–Green–Naghdi equations, which is validated against experimental results for wave transformation over a submerged shoal. Lannes and Marche (2015) offered a new mathematical structure of Green–Naghdi equations which makes them much more suitable for the numerical resolution, in particular in the demanding case of 2DH surfaces.

To close this section, an overall comparison of popular BTMs and weighted residual-type models in terms of various aspects is summarized in the following table.

Models	Phase velocity accuracy (2% error)	Dependent velocity variables	Velocity profile	Basic assumptions	Vorticity	No. of momentum equations	Remarks
1. Fully nonlinear $O(\lambda^2)$ Boussinesq model (Liu et al., 1995; Wei et al., 1995)	3.8	Reference horizontal velocity	u: quadratic w: linear	$O(\lambda^2)$, $\epsilon = 1$	Potential flow (no vorticity allowed)	1	Fully nonlinear extension of weakly nonlinear Boussinesq model (Nwogu, 1993)
2. Fully nonlinear $O(\lambda^2)$ Boussinesq model (FUNWAVE) (Chen, 2006)	3.8	Reference horizontal velocity	u: quadratic w: linear	$O(\lambda^2)$, $\epsilon = 1$	Very weak ($O(\lambda^4)$) horizontal vorticity, leading-order vertical vorticity	1	Derived from Euler equation while keeping vertical vorticity
3. Fully nonlinear $O(\lambda^4)$ Boussinesq model (Gobbi et al., 2000)	7.4	Reference velocity potential	u: fourth w: cubic	$O(\lambda^4)$, $\epsilon = 1$	Potential flow (no vorticity allowed)	1	$O(\lambda^4)$ extension of model 1 but induces high (>3) spatial derivatives
4. Two-layer Boussinesq model (COULWAVE) (Lynett and Liu, 2004a)	7.8	Reference horizontal velocity in each layer	u: quadratic in each layer w: linear in each layer	$O(\lambda^2)$, $\epsilon = 1$ in each layer	Very weak ($O(\lambda^4)$) horizontal vorticity in each layer	2	Two-layer extension of model 1 by maintaining highest spatial derivative = 3

Model		Computational velocity	Velocity profile	Small parameter	Flow		Remarks
5. Extremely dispersive multi-layer Boussinesq model (Liu *et al.*, 2018)	>60	Computational horizontal and vertical velocity within each layer	u: cubic (can go up to degree 5) w: cubic (can go up to degree 5)	No small parameter assumed	Potential flow	8	Extremely accurate in linear and nonlinear properties by using a combined treatment of [Agnon *et al.* (1999); Lynett and Liu (2004a); Madsen and Agnon (2003)], applicable for slowly varying bottom
6. Green and Naghdi (1976a)	1.4	Depth-averaged horizontal velocity	u: uniform w: linear	No small parameter assumed	Restricted vertical profile of horizontal vorticity	1	Can be considered as fully nonlinear extension of Boussinesq model (Peregrine, 1967)
7. Castro and Lannes (2014)	1.4	Depth-averaged horizontal velocity, "shear velocity"	u: uniform + shear component w: linear + shear component	$\mathcal{O}(\lambda^2)$, $\epsilon = 1$	Relatively stronger ($\mathcal{O}(\lambda)$) horizontal vorticity included	4	Horizontal vorticity is restricted to being weak ($\mathcal{O}(\lambda)$), but profile can be arbitrary
8. Two-layer non-hydrostatic model (SWASH) (Stelling and Zijlema, 2003b)	8.8	Depth-uniform horizontal velocity in each layer	u: uniform in each layer w: linear in each layer	No small parameter used	Same as model 6 in each layer	2	Can be considered as a multi-layer extension of model 6, horizontal velocity is discontinuous at layer interface

(Continued)

(Continued)

Models	Phase velocity accuracy (2% error)	Dependent velocity variables	Velocity profile	Basic assumptions	Vorticity	No. of momentum equations	Remarks
9. GN-level K (Webster et al., 2011)	3.3($K = 2$) 6.8($K = 3$) 11.3($K = 4$) 16.9($K = 5$)	Coefficients in velocity polynomial	u: degree ($K - 1$) w: degree K	Horizontal velocity profile assumed to be a polynomial	Horizontal vorticity profile restricted to a polynomial of degree K	K	No explicit expression for pressure field, tedious resulting equations, wave/current conditions restricted by polynomial assumption
10. GK model (Yang and Liu, 2020)	Same as above	Same as above	Same as above	Same as above	Same as above	K	Explicit pressure expression, simpler model equations than Model 9, wave/current conditions restricted by polynomial assumption

11. SK model (Yang and Liu, 2020)	4.8 ($K = 2$) 10.2 ($K = 3$) 18.8 ($K = 4$) 25.1 ($K = 5$)	Same as above	Same as above	Same as above	K	Improved linear and nonlinear wave properties than Models 9 and 10, wave/current conditions restricted by polynomial assumption
12. C-SK model (Yang and Liu, 2022)	Same as above	u: degree ($K - 1$) + current contribution w: degree K + current contribution	Total horizontal velocity decomposed to include prescribed current	Can consider arbitrary vorticity embedded in prescribed current	K	Extension of Model 11 to include the effects of prescribed steady-state arbitrarily sheared current without increasing unknowns

Chapter 4

Potential Flow Models

The potential flow theory assumes that the fluid is inviscid (frictionless) and irrotational (water particles are not rotating). This leads to the continuity equation (conservation of mass) being written as the Laplace equation, i.e.,

$$\Delta\phi = 0, \tag{4.1}$$

where $\Delta = \partial^2/\partial x^2 + \partial^2/\partial y^2 + \partial^2/\partial z^2$ is the Laplacian operator, ϕ is the velocity potential and the velocity is given by $\mathbf{U} = \nabla\phi$. Equation (4.1) is used as the governing equation for solving the potential flow wave problem. For two-dimensional problems (long-crested waves), the governing equation can also take another form in terms of the stream function.

On the other hand, the governing equation can also be posed as the boundary integral equation of Green's theorem, i.e.,

$$\chi\phi(x_0, y_0, z_0) = \iint_S \left[\phi(x, y, z)\frac{\partial\psi}{\partial\mathbf{n}} - \psi\frac{\partial\phi}{\partial\mathbf{n}}(x, y, z) \right] \mathrm{d}S, \tag{4.2}$$

where (x_0, y_0, z_0) is the position that can be located on or enclosed by the surface boundary S, ψ is an arbitrary function that is continuous in the domain except at finite points and is usually expressed in terms of the distance between (x_0, y_0, z_0) and (x, y, z); $\mathbf{n}$ is the unit vector perpendicular to the surface boundary and points outwards. χ is a constant equal to 4π if (x_0, y_0, z_0) is an interior point or equal to 2π if (x_0, y_0, z_0) is located on the boundary.

The primary advantage of using the velocity potential (or stream function) is that it is a scalar quantity. Therefore, the number of unknowns is reduced compared to the Euler or Navier–Stokes equations, as the velocity vectors can be obtained directly by calculating the gradient of the velocity potential (or stream function). Equation (4.2) can be used to find the solution of velocity on the boundary, which reduces the dimension of the problem of solving Eq. (4.1). However, solving Eq. (4.2) requires extra efforts, such as dealing with singularity problems and the inversion of a full matrix.

The boundary conditions of the problem consist of those on the free surface ($z = \eta$):

$$\frac{\partial \eta}{\partial t} + \frac{\partial \phi}{\partial x}\frac{\partial \eta}{\partial x} + \frac{\partial \phi}{\partial y}\frac{\partial \eta}{\partial y} - \frac{\partial \phi}{\partial z} = 0, \tag{4.3}$$

$$\frac{\partial \phi}{\partial t} + \frac{1}{2}|\nabla \phi|^2 + g\eta + \frac{p_a}{\rho} = 0, \tag{4.4}$$

and on the seabed ($z = -h$):

$$\frac{\partial \phi}{\partial z} + \frac{\partial \phi}{\partial x}\frac{\partial h}{\partial x} + \frac{\partial \phi}{\partial y}\frac{\partial h}{\partial y} = 0. \tag{4.5}$$

Equation (4.3) is called the *kinematic boundary condition*. This boundary condition relates the fluid velocity normal to the surface with the local velocity of the surface and enforces that there is no flow across the interface. Equation (4.4) is the so-called *Bernoulli equation* or *dynamic boundary condition*, which can be derived from the Navier–Stokes equations given the inviscid and irrotational assumptions. This equation provides the relationship between the pressure field and the kinematics. The two equations describing the surface boundary conditions, i.e., Eqs. (4.3) and (4.4), can also be written in terms of the velocity potential at the free surface $\tilde{\phi}$. Consequently, they are identical to the canonical pair derivable from the perspective of the Hamiltonian system (Zakharov, 1968). Equation (4.5) enforces that the seabed is stationary and impermeable. For some cases in which the seabed is mobile, such as tsunamis induced by an underwater earthquake, the time dependency of the seabed location must be taken into consideration.

Many numerical methods have been suggested to solve the wave problem described by Eqs. (4.1)–(4.5). They are usually designed following the same concept, as illustrated in Longuet-Higgins and Cokelet (1976). Given that η and ϕ are known at one time step, their values can be updated in time by using Eqs. (4.3) and (4.4) as the prognostic equations in Eulerian, Lagrangian or mixed Eulerian–Lagrangian schemes. The challenge is to find the gradient of ϕ (velocity) as used in the equations. Therefore, the boundary value problem of Eq. (4.1) or Eq. (4.2) needs to be solved with the given boundary conditions. Hence, the numerical models are often named after the methodology that they employ for solving this boundary value problem. In contrast to the conventional Boussinesq-type models, this type of models usually seek for the solutions of the vertical/normal velocity at the free surface. Therefore, they can be used in scenarios regardless of the water depth. This modeling approach encompasses Higher-Order Spectral (HOS) method and Fully Nonlinear Potential Flow (FNPF) method models, which will be reviewed next.

4.1 Higher-order spectral models

4.1.1 *Introduction*

The Higher-order spectral (HOS, also called High-Order Spectral) models were first introduced by West *et al.* (1987) and Dommermuth and Yue (1987) independently. This method applies the perturbation and Taylor series of the velocity potential at the free surface, as follows:

$$\phi(x, y, z = \eta, t) = \sum_{m=1}^{M} \sum_{n=0}^{M-m} \frac{\eta^n}{n!} \frac{\partial^n}{\partial z^n} \phi^{(m)}(x, y, z = 0, t). \qquad (4.6)$$

The main assumptions are that ϕ and η are small quantities of order $\mathcal{O}(\epsilon)$, where ϵ is an indicator of wave steepness and $\phi^{(m)}$ is of order $\mathcal{O}(\epsilon^m)$. Under these conditions, the expansion in Eq. (4.6) can be convergent. The procedure to find the solution of the boundary value problem requires repetitive conversion of the quantities in the Fourier and physical spaces. With the usage of pseudo-spectrum methods, Fast Fourier Transform (FFT) and aliasing error treatment, this problem can be solved efficiently. Therefore, the HOS model is also sometimes referred to as pseudo-spectrum method. The formulation by West *et al.* (1987) is slightly different, but the resultant surface boundary conditions of Eqs. (4.3) and (4.4) can be truncated at a consistent nonlinear order. Hence, the Hamiltonian system is well conserved. Meanwhile, there are alternative forms similar to the concept of the HOS method, e.g., the so-called Dirichlet–Neumann operator or Spectral Continuation method (Bateman *et al.*, 2001; Craig and Sulem, 1993; Nicholls, 1998). The two methods are in general equivalent concerning the formulations of operator expansion as well as free surface boundary conditions (Schäffer, 2008).

The HOS method adopts the Eulerian approach, therefore, the solutions for free surface elevation and velocity potential must have a single value at each grid point in physical space. The consequence is that it cannot simulate overturning waves. To deal with the onset of wave breaking, the wave surface and velocity potential are usually smoothed with artificial filtering techniques. In addition, the solution to the boundary value problem is often truncated to the user specified order M. Higher-order nonlinear terms of order greater than M are ignored, leading to a truncation error that grows with increasing wave steepness. Since the method has assumed that the wave steepness must be small for achieving the convergence of the expansion in velocity potential, its accuracy is reduced for simulating waves with large steepness. In order to overcome this limitation, some recent works have suggested the Enhanced Spectral Boundary Integral (ESBI) method by solving the boundary value problem of Eq. (4.2) with FFTs (Clamond and Grue, 2001; Fructus *et al.*, 2005; Wang and Ma, 2015). To do so, the boundary integral equation can be reformulated as

$$\iint_{S_0} \frac{\mathcal{V}}{l} \mathrm{d}x \mathrm{d}y = 2\pi \tilde{\phi} + \iint_{S_0} \tilde{\phi} \sqrt{1 + |\nabla_{\mathbf{H}} \eta|^2} \frac{\partial}{\partial \mathbf{n}} \frac{1}{l} \mathrm{d}x \mathrm{d}y, \qquad (4.7)$$

where $\tilde{\phi} = \tilde{\phi}(x, y, t)$ is the velocity potential at the free surface, $\mathcal{V} = \partial \phi / \partial \mathbf{n} \sqrt{1 + |\nabla_{\mathbf{H}} \eta|^2}$ is the local velocity of the free surface (i.e., $\mathcal{V} = \partial \eta / \partial t$), S_0 is the projection of S on the horizontal plane, l is the distance between the field point (x_0, y_0, z_0) and source point

(x, y, z), and $\nabla_{\mathbf{H}}$ is the horizontal gradient operator. The surface boundary conditions of the canonical form (Zakharov, 1968) are then rewritten into a skew-symmetric form and are used as the prognostic equations. The evaluation of $\mathcal{V}$ is divided into a convolution part and an integration part, of which the latter is a small term and is only included when the wave steepness becomes large.

For simulating Stokes waves of steepness $kA = 0.3$ (k and A are the characteristic wavenumber and amplitude, respectively) propagating for 1,000 wave periods with periodic boundary conditions, the HOS method needs to set $M = 7$ (Bonnefoy *et al.*, 2010) to achieve the same accuracy as the ESBI (Wang and Ma, 2015). The user of HOS usually requires to perform a sensitivity analysis, i.e., a series of simulations with a number of choices for M until a convergent solution is achieved. Meanwhile, the disadvantage of using the ESBI is that another set of the boundary integral equations needs to be solved in order to obtain the wave kinematics at interior points. However, HOS can utilize the solution for the $\Phi^{(m)}$ obtained to solve Eq. (4.6) at each time step to estimate the wave kinematics beneath the free surface (Bateman *et al.*, 2003). This only applies to the scenarios when the internal kinematics are of concern, such as applications on model coupling with CFD or validation with PIV measurements in laboratory, otherwise the evaluations of wave kinematics at interior points are unnecessary.

4.1.2 *Numerical models and applications*

Due to the usage of FFT, the HOS method is very computationally efficient in comparison with other non-FFT-based fully nonlinear potential flow models. As a result, it is often employed for direct simulation of random directional seas at a large scale ($\sim 10^2$ peak wave lengths) to study the so-called rogue (or freak) waves[1] (Ducrozet *et al.*, 2007; Xiao *et al.*, 2013). The simulation domain can cover a

---

[1]Exceptionally high waves of height larger than twice the significant wave height. The significant wave height is defined as the mean of the largest one-third of waves in a time record.

large region,[2] e.g., of about 128×128 peak wave lengths ($\sim 400\,\text{km}^2$ corresponding to sea states of 10 s peak period in deep sea) as demonstrated in Xiao *et al.* (2013).

The HOS method was originally developed for deep water or constant water depth. The feature for modeling wave propagation over variable topography was introduced later (Fructus and Grue, 2007; Gouin *et al.*, 2014, 2015; Guyenne and Nicholls, 2008; Liu and Yue, 1998; Smith, 1998). The HOS method was also developed to be able to model waves interacting with horizontally linear-shearing currents (Wang *et al.*, 2018b, 2021; Wu, 2004) and vertically shearing currents (Choi, 2009; Guyenne, 2017). The influence of wind on rogue wave formation can also be simulated with a modified air pressure term in Eq. (4.4) following Jeffrey's sheltering mechanism (Kharif *et al.*, 2008; Touboul, 2007). Similarly, the attenuation effects of fragmented ice sheet on waves can also be modeled with a modified pressure term (Guyenne and Pǎrǎu, 2017). Moreover, HOS models can also be used to solve the interactions between waves and submerged body (Liu *et al.*, 1992) or surface-piercing objects (Grue, 2005).

One of the limitations of HOS is that due to the usage of the FFT, the computational domain must be periodic. In most cases the simulations in the periodic domain are started with initial conditions, i.e., imposing an initial free surface elevation and velocity potential. The wave field is allowed to evolve freely after the simulation starts. This imposes a challenge for model validation with laboratory experiments, since in laboratories, the waves are generated from the wavemaker installed along the side(s) of the wave tank. To mimic this wave generation procedure, a study employs a source function or pneumatic wavemaker techniques to excite waves from a source region located inside of the domain, and use a damping zone to absorb outgoing waves near the boundaries of the tank (Clamond *et al.*, 2005). By doing so, the periodical boundary condition is not violated. Alternatively, a non-periodic HOS method has been developed with the capability to model the physical wavemaker motion at one end of the tank, reflection along side walls, and an absorbing zone at the other end (Bonnefoy *et al.*, 2004). Later, this method was

---

[2]Here the term "large" is relative to other phase-resolving models. Such a domain size is still small in comparison with those used in the phase-averaged modeling. For instance, the grid size can be $\sim 10\,\text{km}$ in spectral wave models

improved to account for the second-order (Ducrozet *et al.*, 2006) and third-order (Ducrozet *et al.*, 2012) effects in wavemaker excursion. This HOS model is now freely available as a open-source software known as HOS-Ocean[3] (Ducrozet *et al.*, 2016).

4.2 Fully nonlinear potential flow models

4.2.1 *Boundary integral equation method*

In contrast to the HOS models (which only consider nonlinearities up to the order M), the Fully Nonlinear Potential Flow (FNPF) models solve the full equation of the potential flow theory without truncating the solutions to the boundary value problem. The first attempt was made by Longuet-Higgins and Cokelet (1976) to study two-dimensional overturning waves in deep water. A mixed Eulerian–Lagrangian method was suggested to solve Eq. (4.2), which was later known as the Boundary Integral Equation Method (BIEM) or Boundary Element Method (BEM). This work was subsequently extended to study stability of steep deep water waves (Tanaka, 1983, 1985) and solitary waves (Tanaka, 1986). It also became a popular tool for the study of various types of breakers in finite water depth (Baker *et al.*, 1982; New *et al.*, 1985; Vinje and Brevig, 1981). For more general purposes, the concept of Numerical Wave Tank (NWT) was introduced and many higher-order (in terms of temporal and spatial derivatives) BIEM models were suggested (Dold, 1992; Grilli and Horrillo, 1997; Grilli *et al.*, 1989; Grilli and Subramanya, 1996; Henderson *et al.*, 1999). The higher-order scheme overcomes the numerical instability issues of "sawtooth" appearing on the free surface so that the smoothing procedure is no longer required. These models were successfully applied to simulate wave propagation over varying topography, and interactions between waves and surface-piercing floating objects. The BIEM models were later extended to model three-dimensional wave problems (Boo *et al.*, 1994; Celebi *et al.*, 1998; Ferrant, 1997; Xu and Yue, 1992), and their propagation on varying topography (Grilli *et al.*, 2001; Guyenne *et al.*, 2000; Xue *et al.*, 2001).

---

[3]https://lheea.ec-nantes.fr/valorisation/logiciels-et-brevets/hos-ocean

To further extend the BIEM for practical applications, the BIEM has been used to model the interactions between waves and bubble-structures (Tong, 1997). It can also be used to model different types of waves, such as directional focusing wave (Brandini and Grilli, 2001a,b; Fochesato *et al.*, 2007), tsunami generation by a submarine landslide (Enet, 2006; Grilli *et al.*, 2002), and ship waves (Sung and Grilli, 2005, 2006a,b). Later, the wind effects on waves based on Jeffrey's sheltering mechanism were introduced (Touboul and Kharif, 2010). BIEM can also simulate the wave propagation with ambient steady uniform currents (Ryu *et al.*, 2003), horizontally shearing currents (Moreira and Peregrine, 2012) and vertically shearing currents of constant vorticity, both in deep water (Touboul and Kharif, 2016) and shallow water (Kharif *et al.*, 2017). Some studies also focused on the interactions between waves and floating structures in presence of steady uniform currents (Hermans, 2000; Kim and Kim, 1997; Lin and Hsiao, 1994).

It should be noted that the discretization of the boundary integral formulation together with Green's identities lead to dense non-symmetric matrix operators that cannot be solved in a straightforward way with linear asymptotic scaling (Engsig-Karup *et al.*, 2016). Thus, the BIEM usually costs more computer memory and is less computationally efficient compared to volume-based discretization methods (Ma and Yan, 2009; Wu and Eatock-Taylor, 1994), which will be discussed next.

4.2.2   *Volume-based discretization methods*

The volume-based discretization methods discretize the whole computational domain into a finite number of grid points including the boundaries. This type of model includes the Finite Element Method (FEM), Spectral Element Method and Finite Difference Method (FDM).

The FEM model approximates the solution of the problem in its equivalent integral form using piecewise polynomial shape functions as follows:

$$\phi(x, y, z, t) = \sum_{j=1}^{N} s_j(x, y, z)\phi_j(t), \qquad (4.8)$$

where s_j is the shape function of (x, y, z) associated with the nodal point j, and ϕ_j is the approximated value of ϕ at node j. The solution for ϕ_j can then be obtained by minimizing the error of the approximated solution; such methods are summarized in Zienkiewicz *et al.* (2013). The FEM was firstly suggested for two-dimensional water wave problems in the time domain by Wu and Eatock-Taylor (1995, 1994). It was later extended to model water waves in three-dimensional NWT (Ma and Eatock-Taylor, 1997; Ma, 1998; Wu *et al.*, 1995, 1996). The FEM has been successfully used to simulate interactions between waves with variable water depth and fixed structures (Clauss and Steinhagen, 1999; Sriram *et al.*, 2006; Wang and Khoo, 2005; Westhuis and Andonowati, 1998), floating structures (Hu *et al.*, 2002; Wang and Wu, 2006, 2007) and multi-body floaters (Ma *et al.*, 2001a,b).

One drawback of the FEM is that the mesh needs to be regenerated at every time step, which is time consuming, although some efforts have been made to reduce the CPU time of the mesh generation (Heinze, 2003; Turnbull *et al.*, 2003; Wu and Hu, 2004). The efficiency of the FEM was significantly improved owing to the development of the Quasi Arbitrary Lagrangian–Eulerian FEM (QALE-FEM) model (Ma and Yan, 2006). QALE-FEM was subsequently extended to study wave and floating structure interactions (Yan, 2006) and moored floating structures (Ma and Yan, 2009). Additional features have also been introduced to the QALE-FEM model, such as the capability to deal with overturning waves (Yan and Ma, 2010), effects of steady uniform currents (Yan *et al.*, 2010), effects of wind based on Jeffery's mechanism (Yan and Ma, 2011) and tsunami wave generation and propagation (Yan and Ma, 2014; Yan *et al.*, 2013).

Meanwhile, for the purpose of achieving higher-order accuracy, the Spectral Element Method (SEM) was suggested (Patera, 1984). In contrast to the FEM, the SEM uses trigonometric polynomials, such as Chebyshev or Lagrange polynomials, as the shape function. This method has been widely used in structure analysis while its applications in computational fluid mechanics are summarized in Karniadakis and Sherwin (2013). More recently, the SEM with nodal Lagrange basis functions was introduced for water wave problem in the time domain based on fully nonlinear potential flow theory (Engsig-Karup *et al.*, 2016). There is also current interest in the hybrid Spectral-Finite Element Method that combines SEM's advantages of high accuracy and rapid convergence with FEM's flexibility of handling complex geometries (Robertson and Sherwin, 1999; Xu *et al.*, 2018).

The Finite Difference Method (FDM) has recently received some attention attributed to its high computational efficiency for solving water wave problems based on the fully nonlinear potential flow theory. The FDM was firstly suggested to simulate wave propagation in three-dimensional scenarios accounting for variable water depth (Li and Fleming, 1997). It applies the so-called σ-coordinate transformation in vertical direction:

$$\sigma(x, y, z, t) = \frac{z + h(x, y)}{\eta(x, y, t) + h(x, y)}, \tag{4.9}$$

leading to a regular-shaped computational domain. Then Eq. (4.1) in the new coordinate system (x, y, σ) can be presented in a finite difference form. For example, the second-order derivative can be approximated by a three-point central finite difference scheme, as follows:

$$\frac{\partial^2 \phi}{\partial x^2} \approx \frac{\phi_{m+1,n,l} - 2\phi_{m,n,l} + \phi_{m-1,n,l}}{\Delta x^2}, \tag{4.10}$$

where the subscripts m, n and l are grid point indices corresponding to the x, y and σ axes, and Δx is the grid size. The resulting numerical scheme is claimed to be less memory- and time-consuming than the BIEM.

More recently, an arbitrary-order FDM allowing for a non-uniform grid distribution was proposed for water wave problems in two-dimensional scenarios (Bingham and Zhang, 2007), and was later extended for three-dimensional scenarios (Engsig-Karup *et al.*, 2009). The computational efficiency of the FDM has been further improved using Graphics Processing Unit (GPU) aided parallel computation algorithm (Engsig-Karup *et al.*, 2012). This open source software package is now known as OceanWave3D.[4] However, the application of FDM is limited to wave propagation over variable water depth, and it cannot model waves interacting with floating structures.

---

[4]http://www2.compute.dtu.dk/~apek/OceanWave3D/

4.2.2.1 Limitations

Despite the success of the fully nonlinear potential flow models described in this section, their applications are widely known to be limited to non-breaking waves. In other words, they cannot be employed to simulate wave breaking and water wave problems with strong viscous effects. To overcome this drawback, they are often coupled with Computational Fluid Dynamics (CFD) models based on the Navier–Stokes equations. For example, some software packages offer the solution to couple OceanWave3D with the CFD tool Open-FOAM (Jacobsen, 2017). More details about the coupling between fully nonlinear potential models and Navier–Stokes models will be discussed in Chapter 7.

Chapter 5

Navier–Stokes Models

Navier–Stokes equations are a set of fundamental equations that describe the motion of viscous fluids. The development of this set of highly nonlinear partial differential equations was started by Claude-Louis Navier, who extended the previous work by Leonhard Euler (who developed the so-called Euler equations) to viscous fluids, and was later completed by Sir George Gabriel Stokes in the mid-19th century. For a more complete historical review of the development of Navier–Stokes (NS) equations, the reader is referred to Bistafa (2018). At the end of the 19th century, Osborne Reynolds (Reynolds, 1895) introduced further improvements to the NS equations by decomposing the instantaneous variables into a time-averaged component and a fluctuating component. Reynolds's averaging process led to the Reynolds-averaged Navier–Stokes (RANS) equations, which introduced the effects of turbulence separately, as an additional stress term. Since turbulence is a complex and nonlinear phenomenon, this procedure opened the door to representing the turbulence effects through simplified models.

The general RANS equations can be applied to solve for compressible fluids (e.g., impulsive loading conditions) or stratified flows (e.g., salinity/temperature variations in water depth) and comprise the following conservation of mass equation:

$$\frac{\partial \rho}{\partial t} + \nabla \cdot (\rho \mathbf{U}) = 0, \tag{5.1}$$

where ρ is the density of the fluid, t is time, ∇ is the gradient vector (Hamilton operator) and $\mathbf{U}$ is the Reynolds averaged velocity vector; and the following conservation of momentum equations:

$$\frac{\partial \rho \mathbf{U}}{\partial t} + (\mathbf{U} \cdot \nabla)(\rho \mathbf{U}) = -\nabla p + \rho \mathbf{g} + \nabla \cdot (\mu \nabla \mathbf{U}) - \nabla \cdot (\rho \overline{\mathbf{U}'\mathbf{U}'}), \quad (5.2)$$

where p is total pressure, $\mathbf{g}$ is the vector of the acceleration due to gravity, μ is the molecular dynamic viscosity of the fluid, which is equal to $\rho \nu$ where ν is the molecular kinematic viscosity. The last term on the right-hand side of Eq. (5.2) is called the Reynolds stress. The bar denotes the ensemble average and $\mathbf{U}'$ is the fluctuating component of the velocity vector. This term is highly nonlinear and often requires simplified turbulence models to close the equations, as will be discussed in Section 5.3.4.

The general RANS equations are four and have five unknowns (ρ, $\mathbf{U}$ and P). Therefore, additional equations are needed to close the problem. These can often include the so-called equation of state (EOS, which can have many different forms) and additional conservation laws. For example, in cases in which the fluid density and temperature are linked, the EOS can comprise a coefficient of thermal expansion and conservation of energy may introduce thermodynamic variables such as enthalpy and temperature. In other cases in which density depends on the concentration of a solute (e.g., salt), the EOS may be simply a linear expression, and an advection-diffusion equation will complete the set.

In the cases in which the fluids can be considered incompressible (i.e., fluid density is a constant) the system of equations is directly solvable and the RANS equations can be simplified as follows:

$$\nabla \cdot \mathbf{U} = 0, \quad (5.3)$$

$$\frac{\partial \mathbf{U}}{\partial t} + (\mathbf{U} \cdot \nabla)\mathbf{U} = -\frac{1}{\rho}\nabla p + \mathbf{g} + \nabla \cdot (\nu \nabla \mathbf{U}) - \nabla \cdot (\overline{\mathbf{U}'\mathbf{U}'}). \quad (5.4)$$

Even when the fluid is considered incompressible, the NS equations have been developed with the lowest number of assumptions compared to the remaining approaches described in this book. Consequently, modeling coastal processes with NS equations is the most comprehensive approach and is able to simulate all the known wave transformation processes. However, models based on NS equations are often also highly computationally demanding, restricting their range of applicability in space and time.

This section will introduce three different types of models derived directly from NS equations. Hydrostatic wave models (HWM, Section 5.1) solve a simplified set of equations assuming a hydrostatic pressure distribution. Non-hydrostatic wave models (NHWM, Section 5.2) solve the NS equations, splitting the pressure into its hydrostatic and non-hydrostatic contributions. Finally, Computational Fluid Dynamics models (CFD, Section 5.3) solve the NS/RANS equations directly.

5.1 Hydrostatic wave models

5.1.1 *Introduction*

To solve numerically the full set of NS equations, an iterative procedure is required to find the solution for pressure. Numerical simulation based on the CFD models is computationally expensive, hence the applications of CFD are often limited to local scale of several wave lengths. To model the dynamics of long waves on a regional or global scale, such as tsunamis, tides or storm surges, simplifications to the NS equations must be made. For such long waves, the Mach number (ocean flow speed divided by speed of sound) is not large and their horizontal flow characteristic length is much larger than local water depth. The so-called Boussinesq approximation[1] can be applied, meaning that the water density is assumed to be constant, i.e., $\rho = \rho_0$, except in the buoyancy term (i.e., when multiplied by g) and in the previously discussed equations of state. In addition, a scale analysis shows that in the vertical direction, $\partial p / \partial z$ and $\rho \mathbf{g}$ are the dominant terms in the momentum equation (Eq. (5.4)) (Pedlosky, 1987), and one is ready to apply the following hydrostatic pressure (p_H) approximation:

$$p_\mathrm{H} = p_\mathrm{A} + \rho_0 g (\eta - z) + g \int_z^\eta (\rho - \rho_0)\, \mathrm{d}z, \qquad (5.5)$$

---

[1]The definition for Boussinesq approximation here is to be differentiated from that used in Section 5.3.4.

in which p_A is the atmospheric pressure (i.e., pressure at the free surface level η), ρ and ρ_0 are the density of water and the reference density of water, respectively, and z is the vertical coordinate. This expression can account for stratified flows produced by temperature or salinity differences in seawater.

With the hydrostatic pressure assumption of Eq. (5.5), the RANS equations can be significantly simplified. The so-called Hydrostatic Wave Models (HWM) utilize this assumption. Not to be confused with the Nonlinear Shallow Water Equation (NLSWE) models, the definition by Temam and Ziane (2005) is adopted in this book. In general, the HWM are used in the scenarios where the flow velocity exhibits vertically inhomogeneous structure, so that the discretization in the vertical direction is required. On the other hand, the NLSWE corresponds to the vertically integrated form and does not require vertical discretization. Though it should be noted that the majority of the HWM mentioned in this section support the option to run as an NLSWE model (i.e., the so-called barotropic mode).

5.1.2   *Governing equations*

The governing equations of the HWM are also referred to as the primitive equations (Temam and Ziane, 2005), which consist of the continuity equation and the horizontal momentum equations.

The continuity equation describes the conservation of mass

$$\frac{\partial u}{\partial x} + \frac{\partial v}{\partial y} + \frac{\partial w}{\partial z} = 0, \tag{5.6}$$

which is identical to Eq. (5.3) for incompressible flows with $\mathbf{U} = (u, v, w)$. Integrating Eq. (5.6) from seabed to free surface and applying the kinematic free surface and the bottom boundary conditions, the depth-integrated continuity equation for HWM becomes

$$\frac{\partial \eta}{\partial t} + \frac{\partial}{\partial x} \int_{-h}^{\eta} u \, \mathrm{d}z + \frac{\partial}{\partial y} \int_{-h}^{\eta} v \, \mathrm{d}z = 0. \tag{5.7}$$

The momentum equations after applying the hydrostatic approximation only need to consider the horizontal velocities

(Blumberg and Mellor, 1987)

$$\frac{\partial u}{\partial t} + \frac{\partial uu}{\partial x} + \frac{\partial uv}{\partial y} + \frac{\partial uw}{\partial z}$$

$$= f_c v - g\frac{\partial \eta}{\partial x} - \frac{1}{\rho_0}\frac{\partial p_A}{\partial x} - \frac{g}{\rho_0}\int_z^\eta \frac{\partial \rho}{\partial x}\,dz + D_x, \tag{5.8}$$

$$\frac{\partial v}{\partial t} + \frac{\partial uv}{\partial x} + \frac{\partial vv}{\partial y} + \frac{\partial vw}{\partial z}$$

$$= -f_c u - g\frac{\partial \eta}{\partial y} - \frac{1}{\rho_0}\frac{\partial p_A}{\partial y} - \frac{g}{\rho_0}\int_z^\eta \frac{\partial \rho}{\partial y}\,dz + D_y, \tag{5.9}$$

where $f_c = 2\Omega \sin \Phi$ introduces the Coriolis force with Ω being the angular frequency of Earth and Φ, the geographic latitude, which is important in large simulation domains. D_x and D_y are diffusion terms defined as

$$D_x = \frac{\partial}{\partial x}\left(2\nu_t^h \frac{\partial u}{\partial x}\right) + \frac{\partial}{\partial y}\left(\nu_t^h\left(\frac{\partial u}{\partial y} + \frac{\partial v}{\partial x}\right)\right) + \frac{\partial}{\partial z}\left(\nu_t^v \frac{\partial u}{\partial z}\right),$$
$$\tag{5.10}$$

$$D_y = \frac{\partial}{\partial x}\left(\nu_t^h\left(\frac{\partial u}{\partial y} + \frac{\partial v}{\partial x}\right)\right) + \frac{\partial}{\partial y}\left(2\nu_t^h \frac{\partial v}{\partial y}\right) + \frac{\partial}{\partial z}\left(\nu_t^v \frac{\partial v}{\partial z}\right), \tag{5.11}$$

where ν_t^h and ν_t^v are the horizontal and vertical turbulent eddy viscosity components.

For modeling the ocean circulation, the horizontal flow field may show significant variation in the vertical direction, e.g., discharge of waste or cooling water in lakes and coastal areas, upwelling and downwelling of nutrients, salt intrusion in estuaries, fresh water river discharges in bays and thermal stratification in lakes and seas. In those situations, the continuity equation and momentum equation are often coupled with the conservation equations for temperature and salinity. Using the temperature and salinity, the density is computed according to the equation of state. Its horizontal gradient becomes a forcing term in the momentum equation, which is known as the baroclinic effect. To calculate the advection of temperature and salinity due to the upwelling/downwelling, the vertical velocity w needs be estimated. Because of the hydrostatic approximation, w cannot be obtained prognostically, but must be diagnosed from solving the continuity equation after the horizontal velocities are found.[2]

The problem is closed with the supplement of the turbulence closure equations for the viscosity and diffusivity. This is often used to resolve the fluctuations on turbulent scales due to the usage of relatively coarse computational grids. The explanations of turbulence closure models will be presented in Section 5.3.

The boundary conditions at sea surface and seabed enforce that the internal Reynolds stress be balanced with the wind-induced shear stress at sea surface and bottom friction, respectively:

$$\nu_t^v \frac{\partial \mathbf{U_H}}{\partial z} = \tau_\mathbf{w}, \text{ at } z = \eta, \tag{5.12}$$

$$\nu_t^v \frac{\partial \mathbf{U_H}}{\partial z} = \tau_\mathbf{b}, \text{ at } z = -h, \tag{5.13}$$

where $\mathbf{U_H} = (u, v)$ is horizontal velocity vector, $\tau_\mathbf{w}$ and $\tau_\mathbf{b}$ are wind stress vector and bottom frictional stress vector, respectively.

5.1.3   *Numerical models and applications*

The choice of vertical discretization influences the suitability of a numerical model for a given coastal or ocean region. There are three primary approaches (Griffies *et al.*, 2009):

- *z-coordinate*: The geometric distance below the geoid is discretized as a vertical coordinate (see Fig. 5.1 right panel). It is the most common approach employed among the HWM. The shortcoming of this coordinate is that it can misrepresent the effects of topography (the levels intersect the bathymetry) leading to unrealistic vertical velocities near the seabed. Examples include the earlier versions of MOM[3] (short for Modular Ocean Model) (Bryan,

---

[2]If the fluid is vertically homogeneous, a depth-averaged approach is appropriate. By integrating the momentum equations over the water depth and assuming constant water density, they can be reduced to those used in the Nonlinear Shallow Water Equation (NLSWE) models, i.e., Eqs. (3.7) and (3.8). The vertical velocity w is no longer required for the solution of depth-averaged velocity $(\bar{u}, \bar{v})$ in time domain.

[3]https://www.gfdl.noaa.gov/mom-ocean-model/

1969), UNTRIM[4] (short for Unstructured Tidal, Residual, Inter-tidal Mudflat) (Casulli and Walters, 2000), SUNTANS[5] (short for Stanford Unstructured Nonhydrostatic Terrain-following Adaptive Navier–Stokes Simulator) (Fringer *et al.*, 2006), etc.

- *Terrain-following coordinate*: The vertical distance between sea surface and terrain is rescaled and discretized, such as σ-coordinate (see Fig. 5.1 left panel), etc. It can well represent the topography. However, it is also most well-known for its problem associated with the calculation of horizontal pressure gradient (with error being a function of topography gradient and near-bottom stratification). This technique is employed, e.g., by POM[6] (short for Princeton Ocean Model) (Blumberg and Mellor, 1987), etc.
- *Isopycnal-coordinate*: The vertical discretization is based on the density stratification. This type of coordinate works well for modeling tracer transport, which tends to be along surfaces of constant density. However, because cross-isopycnal mixing is not allowed, it has limited applicability in coastal regions and in the surface and bottom boundary layers. This is used in the Miami Isopycnal Coordinate Model (MICOM) (Bleck and Smith, 1990), etc.

Owing to their advantages and disadvantages, there is usually no optimal vertical coordinate for universal utilities. Hence, some HWMs employ hybrid coordinates model that combines two or more conventional coordinates as described above. For example, the HYCOM[7] (HYbrid Coordinate Ocean Model) (Wallcraft *et al.*, 2009) combines the three approaches described above, specifically, with σ-coordinate in shallow water regions, z-coordinate in mixed layer regions and *isopycnal*-coordinate in stratified regions. Some HWMs (such as HYCOM and MOM6) also adopt the Arbitrary Lagrangian Eulerian (ALE) technique to re-map the vertical coordinate and maintain different coordinates within the domain, leading to increased model time step and reduced spurious numerical mixing (Fox-Kemper *et al.*, 2019).

---

[4]https://wiki.baw.de/en/index.php/UNTRIM
[5]https://github.com/ofringer/suntans
[6]http://www.ccpo.odu.edu/POMWEB/
[7]https://www.hycom.org/

Table 5.1   Summary of the HWMs.

Model name	Numerical scheme	Discretization Vertical	Discretization Horizontal
ADCIRC (v44)	FEM	TF	Tri
CROCO (v1)	FDM	TF	Curvilinear
Delft3D (v3)	FDM	Z/TF	Curvilinear
Delft3D FM (v0)	FVM	Z/TF	Tri-Pen-Hex
FESOM (v2)	FEM/FVM	Z	Tri
FVCOM (v3)	FVM	TF	Tri
HYCOM (v2)	FVM	Z-TF-ISO	Curvilinear
ICON-O	FEM	Z	Tri
MICOM	FDM	ISO	Rectilinear
MIKE3	FVM	Z-TF	Tri
MITgcm	FVM	Z	Curvilinear
MOM6	FVM	Z-TF-ISO	Rectilinear
MPAS-Ocean (v6)	FVM	Z/TF/ISO	Tri
NEMO (v4)	FDM	Z/TF/ISO	Curvilinear
POM (v2k)	FDM	TF	Curvilinear
PSOM (v1)	FVM	TF	Curvilinear
ROMS (v3)	FDM	TF	Curvilinear
SCHISM (v5)	FEM	Z-TF	Tri-Quad
SLIM (v3)	FEM	Z-TF	Tri
SUNTANS	FVM	Z	Tri
UNTRIM (v1)	FDM-FVM	Z	Tri-Quad

Note: "-" denotes a hybrid approach; "/" denotes an optional approach Z: z-coordinate; TF, terrain-following coordinate; ISO, isopycnal-coordinate; Tri, triangular; Quad, quadrilateral; Pen, pentagon; Hex, hexagon.

In terms of horizontal discretization of the computational domain, the irregularities of the coastline or riverbank can introduce significant discretization errors. Some models, as ROMS[8] (short for Regional Ocean Modeling System) (Shchepetkin and McWilliams, 2003), employ orthogonal curvilinear coordinates to align the boundaries of the computational domain with the coastline or riverbank. Alternatively, other models are developed to work with unstructured meshes in order to achieve a better representation of the complex coastline or riverbank. Table 5.1 summarizes some well-known HWMs by their discretization and numerical solution schemes.

---

[8]https://www.myroms.org/

The applications of the HWMs often focus on the ocean circulation processes on global and regional scales. They can now be coupled with sea ice, ice shelves/icebergs and high-resolution atmospheric models, as well as spectral wave models to gain insight of the global climate. Some of the HWMs are considered in the Intergovernmental Panel on Climate Change (IPCC) report, e.g., the MOM6 is employed by Geophysical Fluid Dynamics Laboratory (GFDL). A good review on the applications of the HWMs is given by Griffies *et al.* (2009) and Fox-Kemper *et al.* (2019), so details will not be repeated here.

5.1.4 *Limitations*

It should be noted that due to the hydrostatic approximation, the HWMs are only valid for scenarios when the aspect ratio of the motions are shallow. The global and regional circulation of the ocean (on a scale of $10 \sim 10^3$ km) can be accurately described by the HWMs. They presumably begin to break down on the scale to about 1 km $\sim$ 10 km, where the horizontal scale of the motion is comparable with the vertical scale. Besides, they cannot describe the wind- and buoyancy-driven turbulence in the ocean surface mixed layers (on a scale of <1 km), which are essentially non-hydrostatic (Marshall *et al.*, 1997), and the oceanic boundary layer Large Eddy Simulations cannot use the hydrostatic approximation (Fox-Kemper *et al.*, 2019). A few ocean circulation models presently have the non-hydrostatic capability, and this type of models are often used in relatively small-scale phenomena in the ocean, which will be reviewed next.

5.2 Non-hydrostatic wave models

5.2.1 *Introduction*

Non-hydrostatic wave models (NHWM) have a similar range of applicability to hydrostatic wave models (HWM), being able to simulate cases with large horizontal and vertical domains. Nevertheless, NHWM offer more flexibility, since they do not assume that the pressure distribution is hydrostatic. As a result, NHWM are able to simulate flows in which obstacles (e.g. structures) or buoyancy effects produce significant vertical accelerations locally. Moreover, the non-hydrostatic assumption allows the numerical models to solve nonlinear waves and wave–current interactions, and to model (but not necessarily simulate) wave breaking more effectively (Smit *et al.*, 2010). The practical differences in performance between HWM and NHWM can be observed in Zhang *et al.* (2014a). Their work shows how NHWM is able to simulate nonlinear waves more accurately, and develop flow separation/recirculation behind complex bathymetry features, unlike HWM.

5.2.2   *Governing equations*

NHWM solve the NS equations (Eqs. (5.3) and (5.4)). Consequently, and unlike HWM, the NHWM solve the three momentum conservation equations, as they include the vertical momentum equation. Besides that, and unlike CFD models (introduced in the next section), since most NHWM have been developed as extensions to shallow water models (Ullmann, 2008), they often split the total pressure ($p = p_{\mathrm{H}} + q$) into its hydrostatic (p_{H}) and non-hydrostatic (q) components in the equations, as previously shown in Eq. (5.5).

A practical implication linked to the fact that the horizontal spatial scale is typically much larger than the vertical scale in NHWM is that the difference between the horizontal and vertical scales of turbulence may be very wide, too. As a result, NHWM separate the horizontal and vertical eddy viscosities to account for the non-isotropic turbulent dissipation that "may produce a large difference between horizontal and vertical eddy viscosity coefficients" (SWASH Team, 2020), especially in shallow water conditions. The vertical component of viscosity controls the vertical mixing (i.e. vertical momentum exchange) and is mainly produced by vertical velocity gradients induced by bottom and wind stresses. The horizontal component of viscosity controls the horizontal mixing, which is often important near coastal structures, where "horizontal large-scale eddies generated by lateral shear may have a significant role in horizontal mixing" (SWASH Team, 2020). For further information on which turbulence models can be applied in NHWM to calculate the eddy viscosity, the reader is referred to Cea and Vázquez-Cendón (2007).

With all these particularities in perspective, the governing equations for NHWM are the conservation of mass equation (Eq. (5.6)) and the conservation of momentum equations, which after applying Eq. (5.5) are as follows:

$$\frac{\partial u}{\partial t} + \frac{\partial uu}{\partial x} + \frac{\partial uv}{\partial y} + \frac{\partial uw}{\partial z}$$

$$= f_c v - \frac{1}{\rho_0}\frac{\partial q}{\partial x} - g\frac{\partial \eta}{\partial x} - \frac{1}{\rho_0}\frac{\partial p_A}{\partial x} - \frac{g}{\rho_0}\int_z^{\eta}\frac{\partial \rho}{\partial x}\,dz + D_x, \qquad (5.14)$$

$$\frac{\partial v}{\partial t} + \frac{\partial uv}{\partial x} + \frac{\partial vv}{\partial y} + \frac{\partial vw}{\partial z}$$

$$= -f_c u - \frac{1}{\rho_0}\frac{\partial q}{\partial y} - g\frac{\partial \eta}{\partial y} - \frac{1}{\rho_0}\frac{\partial p_A}{\partial y} - \frac{g}{\rho_0}\int_z^{\eta}\frac{\partial \rho}{\partial y}\,dz + D_y, \qquad (5.15)$$

$$\frac{\partial w}{\partial t} + \frac{\partial uw}{\partial x} + \frac{\partial vw}{\partial y} + \frac{\partial ww}{\partial z}$$

$$= -\frac{1}{\rho_0}\frac{\partial q}{\partial z} + D_z, \qquad (5.16)$$

where D_x and D_y are defined by Eqs. (5.10) and (5.11), respectively, D_z is the diffusion term defined as,

$$D_z = \frac{\partial}{\partial x}\left(\nu_t^h \frac{\partial w}{\partial x}\right) + \frac{\partial}{\partial y}\left(\nu_t^h \frac{\partial w}{\partial y}\right) + \frac{\partial}{\partial z}\left(\nu_t^v \frac{\partial w}{\partial z}\right). \qquad (5.17)$$

After reviewing the governing equations of NHWM, it is easier to see the differences between NHWM and Computational Fluid Dynamics (CFD) models, introduced in the next section. To start with, CFD models are governed by Eqs. (5.3) and (5.4), in which the eddy viscosity is not split into different horizontal and vertical components. Moreover, CFD models solve the total pressure, instead of decomposing it into the hydrostatic and non-hydrostatic components. NHWM are often used to simulate extensive areas reaching hundreds of square kilometers, while CFD models are limited to a few square kilometers, as will later be discussed. There are several reasons why this occurs. First, the mesh with which the space is discretized is

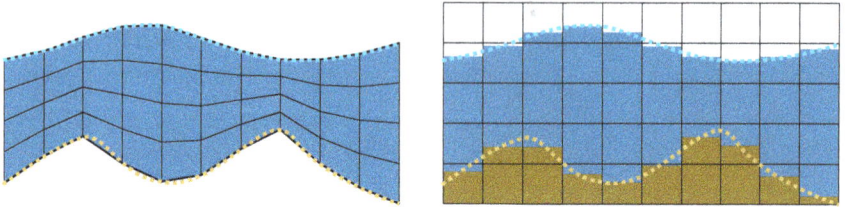

Fig. 5.1   Sketch of typical grids used in NHWM. Sigma-grid (left panel) and
z-grid (right panel). Actual free surface and bottom bathymetry shown in dotted
line. Figure after Ullmann (2008).

usually simpler in NHWM than in CFD. Often NHWM adopt sigma
grids, which are bottom- and surface-fitted, although some models
have the option to use a regular z-grid, in which the free surface
elevation and the cells blocked by an obstacle need to be tracked.
Figure 5.1 shows how these different type of meshes look. The reader
is referred to Ullmann (2008) for a complete discussion of the pros
and cons of each approach.

Second, regardless of the mesh type used, the cell resolution is
lower in the NHWM as compared to CFD. Generally, just a few
cells ("layers") are needed to discretize the whole water column in
NHWM, while CFD requires a significant number to capture the
velocity profile in detail. This, linked with the simpler meshes, limits
the applicability of NHWM, which cannot solve wave breaking (only
model it) because they cannot represent overturning waves. Moreover, NHWM simulations often involve a single phase only. Therefore,
even if wave breaking could be simulated, the air entrainment could
not be reproduced. Consequently, since NHWM can be viewed as a
simplified modeling approach (as compared to CFD), they typically
also use less sophisticated and simpler turbulence models than CFD.
In short, all these elements make NHWM a convenient framework
to simulate wave hydrodynamics in a regional scale at an affordable
computational cost and in a less restricted way than HWM. Yet,
CFD is the most comprehensive approach to simulate waves hydrodynamics in great detail, albeit at a much higher computational cost.

Finally, NHWM can have other equations built in to solve
additional physics complementary to the wave hydrodynamics.
For example, these may be advection-diffusion equations to model

the dispersion of pollutants or sediment transport, as referenced in the next section. The methodologies available to develop couplings of this kind will be discussed in Chapter 7.

5.2.3 *Numerical models and applications*

Non-hydrostatic wave modeling started to be widely applied in the early 2000s. Koçyigit *et al.* (2002) established a NHWM based on a sigma-grid and staggered velocities (i.e., velocities defined at the faces of the cells rather than at the cell centers). The equations were discretized with finite differences and solved with a fractional step method. The model can also run under the hydrostatic assumption. However, the non-hydrostatic mode shows a consistently higher degree of agreement in the challenging cases tested, which include reproducing wave kinematics and wind-driven circulation in a closed tank. Chen (2003) presented a similar model, but developed in Cartesian coordinates and solved the equations with two predictor-corrector steps, assuming hydrostatic pressure in the first of them, and extending the pressure distribution to non-hydrostatic in the second. The model was validated with standing waves, a lock exchange and nonlinear wave transformation over a bar, presenting a reasonable agreement. Nevertheless, the results appear to be sub-par compared to Koçyigit *et al.* (2002). Another example of NHWM, also with excellent results, is Casulli and Zanolli (2002). Their model introduces unstructured grids (i.e. formed by triangular elements), which offer flexibility to fit complex boundaries (coastlines) and the ease to produce mesh refinement and enhance the flow details locally. These features are key to simulating the real case of tides propagating in the Venice lagoon, which being open to the Adriatic sea includes several riverine inlets and narrow channels.

More recent examples of programs that are arguably more established include NHWAVE, SWASH and MIKE3.

The NHWAVE[9] model was presented in Ma *et al.* (2012), and showed an excellent performance in simulating the propagation of a solitary wave, the nonlinear evolution of waves passing over

---

[9]https://sites.google.com/site/gangfma/nhwave

a submerged bar (Beji and Battjes, 1993) and an elliptic shoal (Berkhoff *et al.*, 1982). The model is also capable of simulating wave breaking and run-up, tsunami wave generation by underwater landslides and the development of longshore currents with an outstanding accuracy.

The numerical model SWASH[10] (Zijlema *et al.*, 2011) is widely used in both research and consultancy settings for predicting the transformation of dispersive surface waves propagating from offshore waters to the beach. The model has been applied to study the surf zone and swash zone dynamics, wave propagation and agitation in ports and harbors, as well as rapidly varied shallow water flows typically found in coastal flooding and resulting from e.g. dike breaks, tsunamis and flood waves. SWASH has been developed based on the work of Stelling and Zijlema (2003a), Stelling and Duinmeijer (2003) and Zijlema and Stelling (2005, 2008). A particular feature of SWASH is that it can be either run in depth-averaged mode or multilayer mode, in which the computational domain is divided into a fixed number of vertical layers following the sigma mesh. The frequency dispersion can be improved by increasing the number of layers. Nevertheless, good linear frequency dispersion (1% error in phase velocity) up to $kh \leq 7$ can be achieved with just two equidistant layers.

SWASH is generally more robust and faster than any other Boussinesq-type wave models, since it contains at most second-order spatial derivatives, while the applied finite difference approximations are at most second-order accurate in both time and space. In addition, SWASH does not include any other *ad hoc* models like the surface roller model for wave breaking, the slot technique for moving shoreline or the adaptation of the governing equations to model wave–current interaction. Numerous applications have been reproduced using SWASH. For example, numerical studies on the effects of beach nourishment or breakwaters on wave overtopping were studied in Suzuki *et al.* (2012) and Salas Pérez (2014), respectively. Wave interactions with breakwaters (Al Saady, 2014), mangroves (Khanh, 2019), fringing reefs (Risandi *et al.*, 2020; Zijlema, 2012), permeable pile groins (Zhang and Stive, 2019), ships (Rijnsdorp and Zijlema, 2016) and moored wave energy devices (Tom *et al.*, 2019) have also

---

[10]https://swash.sourceforge.io/

been explored. In addition, field studies of wave run-up on beaches in Yucatan (Mexico) and in Tramanda Beach (Rio Grande do Sul, Brazil) were simulated by Brinkkemper *et al.* (2013) and Guimaraes *et al.* (2015), respectively. SWASH can also be applied at a larger scale. For example, Gracia Garcia *et al.* (2014) incorporated SWASH into a new generation of early warning systems for coastal risk in the North–West Mediterranean Sea.

MIKE3[11] (DHI, 2020a) is a commercial model that has hydrostatic and non-hydrostatic formulations, and is validated in DHI (2020b) for an internal wave seiche, lock exchange problem, open channel flow, density-driven plume dispersion. MIKE3 is completely established and widely used in consultancy (rather than in research), where it is often used for "design, operation and maintenance tasks within the marine environment." However, some validation works, often carried out performing hindcasts, exist (e.g., Passenko *et al.*, 2008).

Many of the HWM have been upgraded to NHWM, extending their range of applicability. For example, Delft3D (Bijvelds, 2001), SUNTANS (Fringer *et al.*, 2006) or ROMS (Ho, 2019). FVCOM also has a version that accounts for a non-hydrostatic pressure distribution (FVCOM-NH, Lai *et al.*, 2010) and XBeach[12] also has a non-hydrostatic module (XBeach-NH) described and validated in Smit *et al.* (2010) and most recently used to simulate run-up in Rutten *et al.* (2021).

Some NHWM can be coupled with other equations or models to simulate additional physics complementary to wave hydrodynamics. Although couplings will be discussed in detail in Chapter 6, some examples are as follows. NHWM can be used to simulate wave–structure interactions in a simplified way. For instance, SWASH has been extended to simulate wave interaction with floating (static/restraint) obstacles such as ships and pontoons in Rijnsdorp and Zijlema (2016) and submerged wave-energy converters (Rijnsdorp *et al.*, 2018). Non-hydrostatic modeling can also be used to simulate sediment transport, for example using XBeach-NH (Mancini *et al.*, 2021) or MIKE3 (Fairley *et al.*, 2015). The latter work shows a real life application, studying the impact that tidal

---

[11] https://www.mikepoweredbydhi.com/products/mike-21-3
[12] https://oss.deltares.nl/web/xbeach/

current turbine arrays have on the sediment at the bottom, and how the sediment in suspension travels afterwards. Another area in which NHWM are used is environmental flows. For example, the non-hydrostatic version of ROMS was used in Ho (2019) to simulate pollutant transport (wastewater effluent plumes).

5.3   Computational fluid dynamics models

5.3.1   *Introduction*

Computational fluid dynamics (CFD) models are currently the most advanced numerical models to simulate the dynamics of fluid flows. CFD models were originally developed as a "universal" approach for flow modeling, and not to solve coastal engineering applications specifically. Therefore, the numerical models and techniques used in the coastal/offshore field are not as mature as in other fields in which CFD has been applied for much longer period of time (e.g., aerodynamics). However, the evolution of computational power, which is cheaper and more accessible than ever before, and the modeling developments in the last three decades have widened the range of applicability and usage rate of CFD models in coastal engineering.

NS equations are derived with a small number of basic assumptions as a starting point, being that the flow is a continuum (and not made up of molecules) and that the fields that describe the flow (pressure, velocity, density...) are differentiable. After posing these assumptions the equations are developed for the flow to obey the universal laws of conservation of mass and momentum. Consequently, CFD models can simulate all the macroscopic physics of the flow. In a coastal engineering context, this implies that NS equations can reproduce complex processes such as wave propagation at all relative water depths, diffraction or even wave breaking in a physical way, without a need for simplified models. Regarding this last point, even though turbulent flow can be simulated with NS equations, it is impractical for most engineering applications and turbulence effects are often modeled instead, as will be discussed later.

As much as CFD modeling is the most comprehensive modeling approach, it is also the most computationally demanding. The high computational costs limit the extension and total time that users can afford to simulate, especially in three-dimensions (3D), which is in the order of magnitude of a few square kilometers and a few hours, respectively. However, hybrid models coupling simpler models and NS models can relax the limitations and extend the range of applicability, as will be discussed in Chapter 7.2.

This section on NS modeling is structured as follows. First, the basic flow equations will be discussed. Then, the different procedures to simulate the free surface in NS equations will be reviewed. The simulation and modeling of turbulence will be introduced next. Finally, the most relevant numerical models available will be outlined along with the most relevant literature on their applications.

5.3.2 *Governing equations*

This part will focus on several topics related to the equations solved by CFD software. First, the NS equations will be described briefly, stressing additional terms that can be introduced for specific purposes. Next, the transformation of NS into a set of equations capable of representing porous materials, ubiquitous in coastal structures, will be introduced.

As already discussed, the CFD models solve the NS or the Reynolds-Averaged Navier–Stokes (RANS) equations, already presented in Eqs. (5.3)–(5.4). The high computational demand for CFD models derives from the fact that this set of equations is highly non-linear and pressure and velocity are strongly coupled.

On top of the standard Eq. (5.4), additional forcing terms can be added to the right-hand side of the equation to account for additional physics and external forces applied to the fluid. A few examples are the surface tension at the free surface of a fluid (Brackbill *et al.*, 1992; Popinet, 2018; Vachaparambil and Einarsrud, 2019) or the force from an obstacle, e.g. introduced by the Immersed Boundary Method (IBM) (Mittal and Iaccarino, 2005), which will be discussed later in this section. Another common type of external forces that can be included in NS equations are friction forces, used to model elements that cannot be solved by the numerical model, either because they are too complex (porous coastal structures, discussed next) or have elements of size below the mesh resolution (vegetation, discussed at the end of this section).

Coastal structures are often built from one or several layers of concrete units or rocks, with the intent to dissipate wave energy through wave breaking and diverting the flow between the solids (viscous and turbulent waves). To simulate a structure like a breakwater in CFD, the flow around each of these elements can be solved if they are explicitly included in the model's mesh (direct approach). However, except under specific circumstances, the concrete units are placed with a random arrangement. Therefore, it is not straightforward to know the position of all the units within a structure (especially when these are in the inner layers) or the geometry of every quarry stone, making it impractical to simulate all the individual elements. Consequently, the direct approach is often limited to representing the largest and simplest elements of the structure only (Altomare *et al.*, 2014; Dentale *et al.*, 2014). The other alternative is to assume that the different layers in the breakwater are homogeneous and isotropic porous materials. The so-called averaged approach disregards the internal geometry derived from the individual elements and substitutes it by an equivalent friction force given by a simplified model. This way, even if the geometry of the obstacles is not explicitly defined in the mesh, its effects on the mean flow can be captured. The averaged approach started being developed in the 1960s for the chemical and petroleum engineering field (Slattery, 1967; Whitaker, 1967) and is based on averaging the RANS equations over a control volume. This averaging process yields the Volume-Averaged Reynolds-Averaged Navier–Stokes (VARANS) equations, which nowadays are the most extended approach for simulating coastal structures in CFD modeling (del Jesus *et al.*, 2012; Higuera *et al.*, 2014a; Jensen *et al.*, 2014; Lara *et al.*, 2008; Liu *et al.*, 1999; Troch and De Rouck, 1998). There is a third approach, which is a hybrid between the two previous ones, in which the largest (and simplest) elements in the layers of a breakwater are simulated directly, while the rest (small or hidden units or rocks) are averaged and modeled using the VARANS equations.

For a full description, derivation and literature review on VARANS equations the reader is referred to Higuera (2015), Losada *et al.* (2016) and Higuera (2021). However, it is worth noting here some of the most important implications of VARANS equations. In the derivation process, the porosity of the material (ratio between the volume of the voids and the total volume of the control volume) is introduced into the equations. Moreover, some terms of the new set of equations, describing the effect of the porous medium on the flow, cannot be solved because the geometry of the individual elements in the porous medium has been averaged out. As a result, these terms need to be modeled to close the equations, and this is done by introducing a set of friction forces to account for the averaged out geometry. There are numerous friction formulations available in the literature (Burcharth and Andersen, 1995; van Gent, 1995), most of which are based on the Forcheimer formulation (Forcheimer, 1901) and comprise a linear term (proportional to velocity), a nonlinear term (proportional to velocity squared) and an added mass term:

$$F = -A\mathbf{U} - B|\mathbf{U}|\mathbf{U} - C\,\frac{\partial \rho \mathbf{U}}{\partial t}. \tag{5.18}$$

The magnitude of each of the terms (A, B, C) depends on the physical properties of the porous medium and the flow (porosity, median grain size, viscosity...) but also on calibration variables. A review on all the expressions available in the literature for the friction factors is provided in Losada *et al.* (2016). These calibration variables are meant to enable the user to adjust the magnitude of the friction based on variables that are not easily measurable, as the shape and angulosity of the rocks or how well the pores are connected. The calibration variables are usually fixed based on a best-fit analysis, which is most often performed comparing experimental and numerical modeling results. Examples on the complete calibration procedure are explained in detail in Jensen *et al.* (2014) and Higuera (2015).

5.3.3   *Free surface*

The NS equations are formulated to simulate the evolution of a fluid. However, they are not able to represent a free surface between different fluids on their own. Obviously, accounting for a free surface is a basic requirement for a CFD software to solve coastal and offshore applications, since most of the problems are driven by waves. Free surface can occur in three different settings. First, between two fluids (e.g., water and air, oil and air...), which is the most standard setting, known as two-phase simulations. Second, between a fluid and a void phase, in which one of the fluids is absent, which is known as single-phase simulations. Finally, between an arbitrary number of

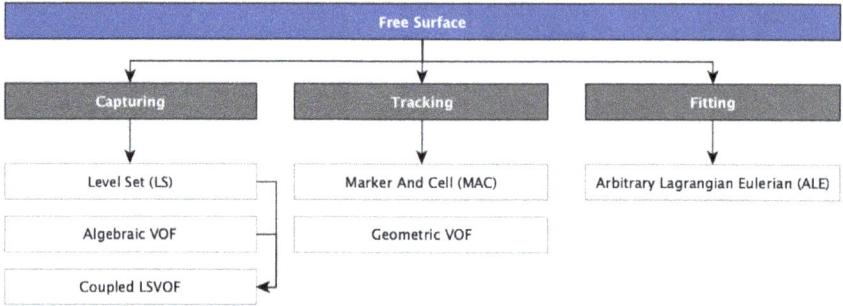

Fig. 5.2   Diagram of the most widely used methods to represent a free surface in NS solvers.

fluids (e.g., water, air, mercury, oil...), which is known as multi-phase simulation. CFD solvers can deal with all three configurations using the techniques outlined in what follows.

There are three main approaches to reproduce the free surface (FS) in CFD solvers, as outlined in Fig. 5.2. These are the FS capturing, FS tracking and FS fitting techniques.

The FS capturing approaches do not specifically track explicitly where the free surface is, but are able to capture/represent it, and do not require the mesh to deform. This group comprises the level set and Volume of Fluid (VOF) approaches. In the level set method (Olsson and Kreiss, 2005; Osher and Sethian, 1988), a signed function (often denoted by Φ) is defined so that each of the cells in the mesh is set to a value equal to the distance to the interface (i.e., free surface), and its sign depends on the fluid which the cell contains. Under these conditions, the free surface would be located at the isosurface in which the level set function $\Phi = 0$. The level set evolves in time according to a simple first-order partial differential equation (Hamilton–Jacobi type). After several time steps, the LS function may become distorted and need to be re-initialized, which takes additional computational effort. However, some recent methods such as Li *et al.* (2005) do not require this step. LS is a low computational cost approach, able to describe complex configurations of the free surface (overturning waves, bubbles and droplets...) in a simple way, and capable of producing accurate geometrical characteristics of the FS (e.g., curvature). The main disadvantage of LS is that most implementations do not conserve mass. The other FS capturing method is the Volume of Fluid (VOF) technique (Hirt and Nichols, 1981). VOF is similar to LS in the sense that the free surface is captured with the so-called indicator function (often denoted by α), but this time the function describes the unit volume of a certain fluid inside each of the cells. For example, a cell full of water will have the value $\alpha = 1$ and a cell full of air will have the value $\alpha = 0$. Any cell with a value $0 < \alpha < 1$ holds a mixture between water and air, and will belong to the interface. Under these conditions, the free surface is often described as being located at the isosurface in which the indicator function $\alpha = 0.5$, but this is just a user convention. In VOF, the evolution in time of the α function is also described with a simple advection equation, as follows:

$$\frac{\partial \alpha}{\partial t} + \nabla \cdot (\alpha \mathbf{U}) = 0. \tag{5.19}$$

Nevertheless, advection equations are known to be diffusive in nature, which will cause the interface to smear over an increasing number of cells as the simulation advantage. Since the FS should be as sharp as possible, numerical techniques such as adding counter-diffusion or compression terms (Deshpande *et al.*, 2012) acting at the interface and applying compressive numerical schemes such as CICSAM (Compressive Interface Capturing Scheme for Arbitrary Meshes), HRIC (High Resolution Interface Capturing) or MULES (Multi-Dimensional Limiter for Explicit Solution) (Márquez, 2013; Wacławczyk and Koronowicz, 2008) are often applied to reduce the issue. The VOF technique is also a low computational cost approach that can describe complex configurations of the free surface (over-turning waves, bubbles and droplets...) in a simple way, and most importantly, it conserves mass. This last point is especially interesting. Given the similarity between Eqs. (5.19) and (5.3), Eq. (5.19) can be seen as a second conservation of mass equation. Unfortunately, the VOF technique is not able to produce accurate geometrical characteristics of the FS (e.g., curvature), in contrast with the LS method.

There is a third approach worth including in this review, and this is the coupled level set and VOF (CLSVOF) technique (Ménard *et al.*, 2007; Wang *et al.*, 2009). As indicated by its name, it is a combination of both LS and VOF that presents the advantages of both methods. Hence, it is mass conservative and able to provide accurate geometrical characteristics of the FS, which results in a higher computational cost of the model.

In the FS tracking approaches the location of the free surface is described explicitly, without requiring the mesh to deform. The most widely-used free surface tracking methods are Marker And Cell (MAC) and the geometric VOF.

The MAC approach was originally developed by Harlow and Welch (1965) and is based on setting markers within the cells of the mesh to track the location of the free surface. These markers are massless particles that are initialized in the region filled by the fluid, and driven by the fluid velocities. By tracking the trajectories of the particles, the location of the FS can be obtained. The main challenges behind this approach are the complexity to apply it in 3D due to the large number of particles needed and potential problems when the fluid elongates and the space between the particles grows. Further developments simplified the original formulation, enhancing its range of applicability (Amsden and Harlow, 1970). For a historical perspective of improvements and more recent developments of the MAC technique, the reader is referred to McKee *et al.* (2004).

The other most important FS tracking method is the Geometric Volume Of Fluid (G-VOF), which shares some features with the Algebraic VOF method described before. The Geometric VOF also uses the VOF technique to report the amount of fluid in each of the cells. However, unlike in Algebraic VOF (A-VOF), Geometric VOF tracks the free surface explicitly in each cell, as a function with a predetermined shape (e.g., a plane). This allows to calculate the fluxes of the free surface geometrically, as it moves driven by the flow velocities. The initial Geometric VOF approach was Simple Line Interface Calculation (SLIC, Noh and Woodward (1976)), in which the free surface within a cell is represented by a horizontal or vertical straight line (at a given location relative to the face according to the VOF value of the cell). The Piecewise-Linear Interface Calculation (PLIC, Youngs, 1982) is another method in which the interface is represented by a straight line with a given slope and relative location, calculated based on the value of VOF in the neighboring cells, but not necessarily matching at the faces shared with the neighbors. There are additional techniques which add more complex configurations and aim to match the functions at the faces, for example the Flux Line-segment model for Advection and Interface Reconstruction

(FLAIR, Ashgriz and Poo, 1991) or spline interface reconstruction (López *et al.*, 2004). Most recently another approach called isoAdvector based on reconstructing the free surface with isosurfaces of a given value of α has also been presented in Roenby *et al.* (2016).

Similarly to A-VOF, in G-VOF updating the free surface relies on an advection step. However, in this case advection is performed geometrically, rather than algebraically, as the functions representing the FS are advected to calculate the fluxes. This involves a subsequent process called the reconstruction step, in which the new functions are determined at each interfacial cell. This process does not exist in A-VOF, and requires a significant computational time, but should produce a perfectly sharp and non-diffusive interface. Another complexity in applying G-VOF is generalizing its formulation to arbitrary (i.e., three-dimensional and unstructured) meshes, which may involve calculating the cuts between the functions and arbitrary polyhedral cells. Consequently, some G-VOF methods are only applicable to structured meshes presently.

The last type in Fig. 5.2 is free surface fitting, which is less widely used than the other types. The main method here is Arbitrary Lagrangian-Eulerian (ALE), in which the free surface is described by a boundary of the mesh (Souli and Zolesio, 2001). Therefore, ALE requires the mesh to move/deform, which is why this method is also known as the boundary-fitted moving mesh method. In ALE, the Eulerian and Lagrangian representations of the flow are blended. The Eulerian description of the fluid focuses on how the fluid flows in and out a control volume (fixed reference frame), whereas the Lagrangian description follows the flow within a certain control volume (moving reference frame). The combination of both achieves a moving boundary that follows the FS, while the internal mesh also moves and is updated to preserve the mesh quality.

The ALE method was originally developed by Noh (1963) and Hirt *et al.* (1974) and its flexibility has made it very useful for problems in which fluids/waves interact with structures (fluid–structure interaction) Hughes *et al.* (1981). In fluid–structure interaction simulations, the flow and structural response are solved with different sets of equations but are coupled, as the fluid impact may deform or displace the structure, and the structure movement will produce a response to the fluid. One of the main advantages of ALE is that the free surface is represented as a sharp (non-diffusive) interface (by the mesh boundary) with the appropriate boundary conditions However, the fact that this method requires mesh movement increases its computational cost. Another disadvantage is that ALE cannot model very complex free surface configurations out of the box, as the mesh deformation cannot exceed some limits that will cause the mesh quality to decrease too much. For example, most ALE models can simulate overturning waves, but only before the wave lip impinges onto the water surface, as afterwards the mesh will be extremely deformed and overlap with itself.

All in all, challenges linked with the simulation of the free surface flows in CFD models, outlined in Higuera *et al.* (2018b), still remain. The first one is parasitic currents (also called spurious velocities), which are unrealistically large numerically generated velocities that appear at the interface between two fluids with a large density difference (such as water and air). The main causes of parasitic currents were discussed in Francois *et al.* (2006), being inaccuracies calculating the curvature of the interface and the imbalance between the pressure gradient (pressure jump) across the interface and the surface tension force. These velocities do not usually impact inertia-dominated flows much (Deshpande *et al.*, 2012), but can have a significant impact on the areas of the flow dominated by surface-tension, such as the meniscus that appears at the shoreline. Very recent developments such as the Ghost Fluid Method (Vukcevic *et al.*, 2017) claim to solve the parasitic current by introducing the jump condition at the interface to couple the two phases implicitly. Another challenge also linked to the shoreline meniscus is the dynamic contact angle at the triple point where air, water and the bottom when the shoreline is moving, which is still an area of active research (Afkhami *et al.*, 2009).

5.3.4   *Turbulence*

Another area in which CFD models excel is simulating turbulence. Turbulence is a physical process by which the energy of the flow is dissipated. This occurs by breaking down eddies into smaller ones several times, thus transferring momentum down to smaller and smaller scales progressively, until they are dissipated (converted to heat) by the molecular viscosity of the fluid at the lowest scale (known as the Kolmogorov scale). Turbulence may not always play a significant role in coastal engineering applications. For example, turbulent effects are negligible during wave propagation. However, turbulent effects can be very important when conditions are dissipative, i.e., when wave breaking occurs and when waves interact with the external layers of porous breakwaters. A full review of turbulence in the coastal engineering field can be found in the book by Sumer and Fuhrman (2020).

Turbulence can be simulated directly in CFD by solving the NS equations in what is called the Direct Numerical Simulation (DNS) approach. The NS equations look almost identical to the RANS equations shown in Eqs. (5.3) and (5.4). The only difference is that the last term on the right-hand side of Eq. (5.4) does not appear, because NS equations have not been averaged.

As mentioned above, DNS is able to simulate turbulence (i.e., solve all the processes physically), which involves capturing all the scales down to the Kolmogorov scale in the mesh. The Kolmogorov scale ultimately depends on the Reynolds number of the flow, but may often be several orders of magnitude smaller than 1 mm. As a result, DNS cases of a typical coastal engineering application may require meshes of billions and billions of cells, which would involve such high computational requirements that they are out of reach for the current computational capabilities. This is why the DNS approach is limited to simulating fundamental research applications with relatively small domains and low Reynolds numbers.

Another approach which requires significantly less computational resources than DNS is the Large Eddy Simulation (LES). In LES, the turbulence for the largest eddies (scale larger than the cell size) is resolved, while the smaller eddies (scale smaller than the cell size) is modeled with a so-called subgrid-scale model (SGS model). Consequently, LES allows using a much larger cell size than DNS, yet still significantly small.

The NS equations are filtered spatially in LES, which results in a similar set of equations as Eqs. (5.3) and (5.4), but with a different last term on the right-hand side of Eq. (5.4). This last term is equal to $\nabla \cdot (\rho \boldsymbol{\tau})$, where $\boldsymbol{\tau}$ is the subgrid-scale stress tensor. This term cannot be calculated and needs to be modeled instead. Using the Boussinesq eddy viscosity hypothesis, the SGS tensor can be modeled in the same way as the viscous term (second to last term on the right-hand side of Eq. (5.4)), using a SGS dynamic viscosity (μ_t) to close the problem. Therefore, in practice the viscous term is computed with an effective viscosity (μ_{eff}), which comprises the molecular and turbulent components of viscosity. The most commonly used model to calculate the SGS viscosity is the one-equation model Spalart–Allmaras (Spalart and Allmaras, 1992).

The use of LES is starting to be more widespread because of the recent improvements in computational power. However, LES is still considered a computationally expensive model and more widely used in research rather than in consultancy projects in the coastal engineering field.

The next approach is based on the RANS equations. In the RANS approach, turbulence is not simulated (resolved) but modeled throughout the whole range of scales, which allows using larger cells than in DNS or LES, which decreases the computational costs significantly.

The RANS equations, already introduced in Eqs. (5.3) and (5.4), are time-averaged. Due to this process, the last term on the right-hand side of Eq. (5.4) (the Reynolds stresses) cannot be simulated and needs to be closed (modeled). Two different options exist for this purpose. The first option involves solving for the Reynolds stresses, which comprise four components in 2D simulations and six components in 3D simulations. Therefore, this approach, known as Reynolds Stress Model (RSM), often involves solving a set of 5–7 additional partial differential equations. Alternatively, the other approach is using an Eddy Viscosity Model (EVM), which by applying the Boussinesq approximation[13] assumes that the Reynolds stresses can be modeled in the same way as the viscous term, with a turbulent dynamic viscosity (μ_t), also called eddy viscosity. Thus, similarly to the LES case, the viscous term is computed with an effective viscosity (μ_{eff}), which comprises the molecular and turbulent components of viscosity. The EVM approach can be implemented with different turbulence models of different complexities, from two-equation models (e.g., k–ϵ, k–ω, k–ω SST), one-equation models (e.g., Baldwin–Barth, Spalart–Allmaras) to algebraic models (e.g., Cebeci-Smith, Baldwin-Lomax).

---

[13]To be differentiated from the definition in Section 5.2.

EVM often provides an accurate representation of the turbulent effects, but this is limited when turbulence is isotropic, which is a main assumption of Boussinesq approximation. In cases in which turbulence is highly anisotropic, RSM will performs better but will also increase the computational cost. To combine the advantages of both approaches, most recently a blended version between RSM and EVM has been developed by Wang *et al.* (2020b).

RANS is the most used turbulence modeling approach presently, both in the research and in the consultancy fields in coastal engineering, and within RANS, the two-equation EVMs. However, one of the main shortcomings that the main two-equation EVMs present is an overproduction of turbulence which dissipates too much energy and damps wave heights artificially in long simulations. The reasons behind this behavior are that an excess of turbulence is spuriously generated at the interface between water and air and that some RANS turbulence models are unconditionally unstable and produce turbulence even when waves are propagating in a potential flow regime. Both issues have been solved in recent works by Devolder *et al.* (2017) and Larsen and Fuhrman (2018).

The final turbulence modeling approach worth mentioning in this review is called Detached Eddy Simulation (DES) and is a hybrid between the LES and RANS approaches. DES uses the RANS approach to model turbulence at the areas in which applying LES is computationally costly because cells would need to be too small, which is usually the case near the walls. In the rest of the domain, turbulence is still modeled using LES, which provides a more sophisticated description of the turbulence effects. As a result, the DES approach provides a good balance between the LES and RANS performance at an average cost.

5.3.5 *Numerical models and applications*

CFD software able to solve the free surface can be tracked back to the 1980s (Nichols *et al.*, 1980). Historically, the first CFD solvers were two-dimensional in the vertical plane (2DV), as the computational cost of fully three-dimensional (3D) models was out of reach for the computational resources at the time.

The adaption in coastal engineering took some time and the first CFD modeling works in the field started appearing only around the year 2000. Some of the earliest works were performed with the models COBRAS (Lin and Liu, 1998) and VOFbreak (Troch and De Rouck, 1998), simulating wave breaking and wave interaction with coastal structures, respectively.

CFD models have incorporated new features and enhanced existing ones. There are two features which were key to generalize the use of CFD in the coastal and offshore fields: wave generation and absorption and flow through porous media. Since waves are the main driving dynamic in most coastal engineering studies, wave generation needs to be as accurate as possible to prevent the propagation of errors. After waves are generated, they will propagate and interact with elements such as the bathymetry or the structures and may reflect back toward the boundaries. If wave reflections are not handled properly, re-reflections at the boundaries will occur and the energetic level of the simulation will keep increasing. This is why an efficient wave absorption technique is key to producing physical results in numerical simulations. A number of wave generation and absorption techniques have been developed over the years, each with different ranges of applicability, efficiency and computational costs. These techniques, which range from relaxation zones to computationally efficient boundary conditions, are reviewed extensively in the chapter by Dimakopoulos and Higuera (2021).

Most of the structures built in the coastal and offshore environments include porous materials, which aid to protect the structure and dissipate wave energy. Therefore, treating the flow within these materials in the most physical way while maintaining a manageable computational cost is a significant requirement for CFD models. As was outlined in the previous section, the VARANS approach is widely used for that reason, and numerous numerical models support it. For an extensive review of these, the reader is referred to Losada *et al.* (2016) and Higuera (2021).

The initial CFD models that were introduced in coastal engineering followed the Eulerian approach. This means that the evolution of the flow was calculated based on a fixed mesh or grid, in which the conservation laws (Eqs. (5.3) and (5.4)) are evaluated in fixed control volumes (cells). However, almost at the same time a new type of modeling based on the Lagrangian approach was introduced.

The Lagrangian description of a fluid follows its particles, therefore Lagrangian methods use discrete elements to solve the NS equations and reproduce the flow behavior. Among the Lagrangian approach, two main methods were adopted, the Smooth Particle Hydrodynamics (SPH) method (Monaghan, 1994) and the Moving Particle Semi-implicit (MPS) method (Koshizuka, 1995). Despite being originally developed to solve astrophysics problems (Monaghan, 1992), SPH modeling also took root in coastal engineering and over the years its popularity increased and led to high-impact works such as Dalrymple and Rogers (2006). Presently, the most widely used SPH models are GPUSPH (Dalrymple *et al.*, 2010), DualSPHysics (Gómez-Gesteira *et al.*, 2012; Gomez-Gesteira *et al.*, 2012) and ISPH (Shao, 2006). The main advantage of SPH models is their flexibility to deal with complex geometries, as they are meshless approaches (i.e., they do not require a mesh). However, this type of modeling is in an earlier stage of development compared to Eulerian modeling and still presents challenges when calculating pressure fields. Therefore, Lagrangian modeling is less widely adopted presently. The reader is referred to Gotoh and Khayyer (2018) for an extensive review of the state of the art of particle methods in coastal engineering, and to Lind *et al.* (2020) for the most recent and review on SPH.

During the mid- and late-years of the 2000s decade the CFD models were extensively validated to simulate the interaction between waves (ranging from solitary waves to regular and irregular sea states) and a wide range of coastal structure typologies. The IH2VOF model,[14] formerly called COBRAS-UC (since it is derived from COBRAS, Losada *et al.*, 2009), was one of those models. Lara *et al.* (2006) studied irregular waves interacting with a submerged breakwater, Losada *et al.* (2008) researched overtopping of rubble mound breakwaters and later, Guanche *et al.* (2009) analyzed wave loads and stability of a high-mound and a vertical breakwater. Surf and swash zone hydrodynamics were also studied with this model (Torres-Freyermuth *et al.*, 2010).

In the decade of the 2010s, the 2D models eventually led to 3D models, boosted by the easier access to more powerful computational resources. However, limitations still exist to simulating large domains

---

[14]https://ih2vof.ihcantabria.com

and long time series in 3D due to the high computational costs involved. There are several examples of in-house (research) and commercial (licensed) models worth mentioning. In the research realm, CADMAS-SURF 3D (Arikawa *et al.*, 2007; Okumura, 2014) has been applied to simulate breaking waves and the interaction between waves and submerged breakwaters; IH3VOF (del Jesus *et al.*, 2012; Higuera *et al.*, 2013a; Lara *et al.*, 2012) has been applied to simulate wave breaking and wave interaction with impermeable and porous structures. In the commercial realm, FLOW-3D (Choi *et al.*, 2007; Hemavathi and Manjula, 2019; Jin and Meng, 2011; Ko *et al.*, 2015) has been applied to simulate wave impacts on coastal highway bridges, scour around offshore jackets and attenuation by coastal vegetation. ANSYS FLUENT (Cai *et al.*, 2018; Didier *et al.*, 2016; Park *et al.*, 2018) has been used to develop a 3D numerical wave tank and widely applied to simulate impacts of waves on bridge decks. The list of commercial and research numerical models available is very extensive and not all the applications can be included in this review. Nevertheless, the best-known numerical models with applications in coastal and offshore engineering are included in Appendix.

This decade also experienced the development and boom of open source models. Open source models (OSM) are often multi-purpose, but since they are free and provide the source code (unlike commercial models), they offer the users significant flexibility to adapt the code to suit their specific needs. Other benefits of OSM include comprehensive documentation and an online community that helps accelerate and test the new developments. Similarly to what happened when 2D CFD models were introduced, the initial efforts always focus on developing the basic capabilities (e.g., wave generation and absorption, porous media flow...) and validating those using existing benchmarks.

The most widely used OSM nowadays is OpenFOAM[15] (Weller *et al.*, 1998). The initial works in which OpenFOAM was applied to coastal and offshore engineering involved producing and validating wave generation and absorption libraries. The most relevant examples are waves2FOAM[16] (Jacobsen *et al.*, 2012) and

---

[15]https://openfoam.org, https://openfoam.com
[16]https://openfoamwiki.net/index.php/Contrib/waves2Foam

IHFOAM[17] (Higuera *et al.*, 2013b,c). Most recently, the model olaFlow[18] (Higuera, 2017, 2020) was introduced, providing significant improvements in active wave absorption. Regardless of the wave generation method (a comparison is available in Windt *et al.*, 2019), OpenFOAM has been validated and applied to simulate porous coastal structures (Higuera *et al.*, 2014a,b; Jensen *et al.*, 2014; Lee *et al.*, 2019), to study wave groups and freak waves (Vyzikas *et al.*, 2015), loading on offshore structures (Hirai *et al.*, 2018), wave dissipation by vegetation (Jacobsen and McFall, 2022; Maza *et al.*, 2016), scour under and around pipes and piles (Larsen *et al.*, 2016; Liu and García, 2008) and wave-energy converters (Ding *et al.*, 2019). For all these reasons, OpenFOAM is currently the most versatile and flexible numerical modeling framework for coastal and offshore engineering. Moreover, OpenFOAM is widely established both in research and in consulting, and has even become a standard requirement in some consultancy projects.

Another example of OSM with similar capabilities but significantly lower adoption rate is Reef3D.[19] The numerical model has been applied to simulate extreme waves (Bihs *et al.*, 2019), wave interaction with cylindrical structures (Bihs *et al.*, 2017), with porous breakwaters (Sasikumar *et al.*, 2019).

New OSM possess advanced features that are added and enhanced continuously by the community of developers. These modern solvers are now able to include additional physics complementary to the Navier–Stokes equations, such as porous media flow, floating structures, vegetation dynamics and transport and diffusion of substances. Some of these advanced features are implemented as "add-ons," while others require solving additional sets of equations and should be regarded as couplings between models instead. The latter ones will be the focus of the next section.

The Immersed Boundary Method (IBM) (Mittal and Iaccarino, 2005; Peskin, 2002) is a feature that allows representing very complex obstacles (e.g., solid structures) within the mesh, instead of needing to mesh around them. Mesh generation is often a complex and

---

[17]https://ihfoam.ihcantabria.com
[18]https://olaflow.github.io
[19]https://reef3d.wordpress.com

lengthy step in which users need to ensure that the cells and mesh fulfill adequate quality criteria. The IBM technique blocks those cells inside the obstacle and ensures that the cells at the interface provide an approximation to the no-slip boundary condition. Although this allows a simple representation of obstacles without complex meshes involved, a sufficient mesh resolution is required to resolve the flow correctly and some challenges to solve the boundary layer flow accurately still exist for this approach. IBM also allows simulating moving elements (e.g., piston wavemaker or floating structure) with significant motions that are not possible for deforming meshes. Some examples of practical applications include Shen and Chan (2008), in which IBM is used to simulate how wave propagation is impacted by different solid obstacles located at the sea bottom, and to generate tsunami waves with a moving bottom; Zhao *et al.* (2020), in which IBM is used to reproduce tsunami wave impacts on bridge decks.

The overset mesh or chimera grid approach (Chandar *et al.*, 2018; Windt *et al.*, 2018) allows merging two types of meshes into a single one. The base mesh covers the whole domain and a smaller mesh, contained within the base mesh, is where the structure/obstacle exists and where the flow near the structure is calculated. This secondary mesh is able to undergo large movements with up to 6 degrees of freedom to simulate floating structures, if needed. At the interface between the two meshes there is an exchange of fields (similar to a two-way coupling explained in Chapter 7) so that the whole system continues to be driven by the conservation equations. This step is performed with grid-to-grid interpolation techniques. Overset meshes have been used in ship hydrodynamics and propulsion (Shen *et al.*, 2015) and to simulate wave-energy converters (WEC) subjected to extreme waves (Katsidoniotaki and Göteman, 2020).

5.3.6   *Coupling additional equations*

A way to extend the capabilities of CFD solvers is to add modules that solve additional equations simultaneously to the main NS simulation. This can be done in the same solver (as an "add-on") or by coupling two different models. The basics on coupling techniques will be later described in Chapter 7.

An example on coupling additional equations has already been discussed, and this is the porous media flow (VARANS) equations, which are coupled with the RANS equations inside porous media. The similarity between both sets of equations make the coupling process straightforward.

In other cases, the CFD model can be coupled with solid mechanics equations, such as Newton's second law (i.e., $F = ma$, force equals mass times acceleration), to calculate the response of a body (assumed rigid) subjected to waves. Several examples have already been listed previously, as ship hydrodynamics (Shen *et al.*, 2015) and WECs (Katsidoniotaki and Göteman, 2020) are common types of floating structures and can experience large displacements during the simulation, therefore, they can also benefit from the overset mesh technique.

If the stresses and deformations of the structure are required, the CFD model needs to be coupled with a structural solver. Structural solvers often use the finite element method (FEM) and can be coupled at the same time or asynchronously, as later discussed in Chapter 7. Due to the high computational cost involved, and the complexity in coupling different models, there are not many examples for coastal applications: a hydroelastic solver using the SPH (Khayyer *et al.*, 2020b) and the MPS methods (Khayyer *et al.*, 2020a). Nevertheless, a general framework that could be applied to coastal processes in the future is the OpenFPCI framework (Hewitt *et al.*, 2019), which is open source and based on OpenFOAM. Simpler options also exist for simpler (or idealized) geometries. Chen and Zou (2019) simulate the interaction between waves and vegetation. Vegetation is accounted for as a solid Lagrangian entity, which is able to move and deform driven by the fluid. At the same time, vegetation introduces a drag force on the fluid via IBM. This is a particular example in which the IBM technique can be used to represent the effect of elements that are smaller than the mesh size, and because of that, they cannot be fully represented by the mesh; in this case, for vegetation, which is formed by elements that are long and slender.

Another type of model that can be coupled with CFD is the Discrete Element Method (DEM), often used to simulate the interaction between different solids. This method has been coupled with CFD to simulate how concrete units interlock at the external layers of breakwaters (Latham *et al.*, 2008; Xiang *et al.*, 2012). Sediment transport can also be simulated with the DEM method, either representing individual particles as the solids (if sediment is coarse enough) or parcels of sediment (i.e., a group of individual particles that stay together). Examples of this include the simulation of bed transport, (Sun and Xiao, 2016) and erosion on a beach (Higuera, 2021).

Sediment transport and scour are critical processes, which can lead to the catastrophic failure of coastal and offshore structures, that can also be modeled in CFD. The most usual approach to simulate sediment transport is to treat sediment as a continuum, and not the DEM method previously introduced, which is computationally much more expensive. Within the continuum approach, there are multiple techniques to simulate sediment, ranging from treating the sediment bed as a moving boundary, which moves down when erosion happens (and puts sediment into suspension) and moves up when accretion occurs (trapping sediment that was in suspension). The sediment in suspension can then be simulated with a simple advection equation. This method has been used to simulate the back-filling processes around a circular pile under wave action (Baykal *et al.*, 2017), and later extended to include seepage flow as well (Li *et al.*, 2020). Sediment can also be treated as an additional phase to water and air. In this case, semi-empirical formulations are required to account for the rheology of the sediment, as for example in Ouda and Toorman (2019).

Wave–soil interactions can produce settlement of coastal and off-shore structures due to the consolidation of the soil. This effect, which may eventually lead to structural failure, is caused by the cyclic pressure variation driven by waves, which may build up the pressures on the soil and eventually lead to liquefaction. The first general theory on soil consolidation in three-dimensions was developed by Biot (1941), and since then the so-called Biot's poro-elastic model is widely used in diverse engineering fields. The basic concept behind Biot's theory is that the pore pressure within a (fully or partially) saturated porous medium (i.e., the soil) has a contribution in the total stress tensor. Consequently, changes in the flow pressure will impact the bearing capacity of the soil. For example, a reduction in pore pressure, which can be induced by the troughs of waves, can lead to consolidation, especially when the soil is subjected to significant loading from a structure placed on top of it. For a complete theoretical framework on poro-elasticity, the reader is referred to Guéguen *et al.* (2004).

The applications of poro-elasticity in a coastal engineering framework can be traced back to Madsen (1978), who set the theoretical background for the pore pressures and effective stresses induced in a porous bed by waves. When coastal structures also exist, it becomes a wave-seabed–structure interaction problem, which is highly complex because it would involve coupling three models. Consequently, this type of modeling has been approached in simplified ways, mostly regarding the wave and structural models. For example, the work by Mostafa *et al.* (1999) developed a coupling between a Boundary Element Method (BEM) (to solve the wave hydrodynamics in the clear region assuming potential flow conditions) and a Finite Element Method (FEM), which solves the flow and pressures within the porous media (i.e., the breakwater foundation and soil). The structural component of the hybrid model is simplified, by including it into the porous media model with a very large stiffness and very low permeability. The model was verified against a set of physical experiments, proving to produce satisfactory results in terms of free surface elevation and pressures inside the porous media. Jeng *et al.* (2001) simulated the wave-induced pore pressure around a composite breakwater, considering simplifications such as potential flow theory (incompressible and irrotational flow) for waves and no displacement of the structure. Jeng and Ou (2010) extended their model to three-dimensional domains, to simulate wave-induced pore pressures in the vicinity of breakwater heads. The software considers two different soil response models, poro-elastic and poro-elasto-plastic, to compare the pore pressure variation and liquefaction induced by waves. The poro-elasto-plastic model was able to capture both the oscillatory and residual components of the excess pore pressure, which are crucial to estimate the potential liquefaction hazard. Another conclusion of this work is that "soil characteristics affect the pore pressure slightly more significantly than wave characteristics," therefore, a correct characterization of the soil is important to obtain meaningful results for a particular location. A later work (Jeng and Ye, 2012) extended the 3D analysis to a complete rubble-mound breakwater, focusing on the consolidation of the soil rather than on the liquefaction potential.

Only recently the hydrodynamic modeling in wave–soil interaction models has been extended to solve the NS equations (CFD approach). This approach adds significant computational cost, but offers full flexibility to solve all the wave transformation processes, including wave breaking. Examples of this in the literature include applications for a buried pipeline (Liang and Jeng, 2021) subjected to combined irregular waves and currents in 2D-vertical. Fully 3D results using the same model are presented in Cui *et al.* (2021) for offshore foundations subjected to waves and currents. The results highlight the importance of running simulations in 3D when 3D effects are expected (e.g., wave diffraction, waves and currents with different angles of incidence).

Chapter 6

Lattice–Boltzmann Models

6.1 Introduction

The Lattice–Boltzmann Method (LBM) is a flow modeling approach that is still in an early stage of development in coastal engineering, but that presents promising features. LBM has some parallelisms with the SPH method introduced in Section 5.3.5. Both were originally developed to solve problems that, *a priori*, might seem unrelated with wave modeling, namely SPH for astrophysics (Monaghan, 1992) and LBM based on gas dynamics methods (Lattice Gas Automata, Hardy *et al.*, 1973), and both can replicate the Navier–Stokes equations.

Generally, the description of flows can be separated into three scales. The microscopic scale (i.e., at a molecular level) tracks each individual particle and deals with their interactions. The mesoscopic scale deals with simulating the distribution of particles in a discrete way, and not the individual particles. Finally, the macroscopic scale assumes that the flow is a continuum, as previously explained when introducing the Navier–Stokes equations.

LBM is derived "to construct simplified kinetic models that incorporate the essential physics of microscopic or mesoscopic processes so that the macroscopic averaged properties obey the desired macroscopic equations" (Chen and Doolen, 1998). In other words, the macroscopic dynamics obtained from the LBM are derived directly from the microscopic dynamics. Consequently, LBM is able to retain some fundamental physics inherent to the microscopic dynamics, which are lost in other types of modeling (e.g., NS) due to the underlying assumptions. Therefore, this highlights the LBM suitability to deal with applications such as microfluidics or porous media flow (Guo and Shu, 2013), in which the continuum assumption may be starting to be pushed to its limits.

There are several other advantages of the LBM, as noted in Chen and Doolen (1998). The first advantage is that in the LB kinetic equation (see next section) the convection operator is linear (Perumal and Dass, 2015), unlike in NS equations, in which convection is nonlinear. Secondly, "the pressure of the LBM is calculated using an equation of state" (Chen and Doolen, 1998), which is significantly simpler than solving the pressure Poisson equation in NS equations. Third, LBM considers only a limited set of directions for the velocities in space, which implies that the LBM equations can be solved with simple arithmetical operators.

These features revert in three other practical advantages, namely the ease of implementing boundary conditions and treating complex geometries, and the capability of fully (and massively) parallelizing the solvers.

6.2   Equations

As mentioned earlier, the LBM is derived from the Lattice Gas Automata (Hardy *et al.*, 1973), which is based on the kinetic theory. Krüger *et al.* (2017) indicate that the fundamental variable in kinetic theory is the particle distribution function $f(\mathbf{r}, \boldsymbol{\xi}, t)$, which depends on the position ($\mathbf{r} = (x, y, z)$), particle velocities ($\boldsymbol{\xi} = (\xi_x, \xi_y, \xi_z)$) and time ($t$). The particle distribution function represents the density of particles with a velocity $\boldsymbol{\xi}$, at position $\mathbf{r}$ and time t (Krüger *et al.*, 2017). As a result, the macroscopic properties such as the density (ρ) and the velocity of the fluid ($\mathbf{U}$) can be calculated as

$$\rho(\mathbf{r}, t) = \iiint f(\mathbf{r}, \boldsymbol{\xi}, t) \, \mathrm{d}\boldsymbol{\xi}, \qquad (6.1)$$

$$\rho(\mathbf{r}, t) \, \mathbf{U}(\mathbf{r}, t) = \iiint \boldsymbol{\xi} \, f(\mathbf{r}, \boldsymbol{\xi}, t) \, \mathrm{d}\boldsymbol{\xi}, \qquad (6.2)$$

respectively. Based on Eq. (6.1), the particle distribution function f can be seen as a representation of the fluid density in a seven-dimensional space of position (x, y, z), particle velocities (ξ_x, ξ_y, ξ_z) and time (t).

The evolution of the f function in time is governed by the Boltzmann equation (Krüger *et al.*, 2017):

$$\frac{\partial f}{\partial t} + \boldsymbol{\xi} \cdot \frac{\partial f}{\partial \mathbf{r}} + \frac{\mathbf{F}}{\rho} \cdot \frac{\partial f}{\partial \boldsymbol{\xi}} = \Omega(f), \qquad (6.3)$$

where $\mathbf{F}$ corresponds to the body forces (e.g., gravity) and $\Omega(f)$ is a source term. The first and second terms represent the f function being advected by the particle velocities (ξ_i). The third term represents the forces affecting the particle velocities. The source term represents the "local redistribution of f due to collisions" of the particles, and is called the collision operator.

The Boltzmann equation is more general than NS equations. As a result, not all the solutions of the Boltzmann equation satisfy NS equations (Krüger *et al.*, 2017). However, "the incompressible (NS) equations can be obtained in the nearly incompressible limit of the LBM" (Chen and Doolen, 1998). Therefore, it can be concluded that LBM can solve the NS equations. For the complete derivation of the macroscopic conservation equations (i.e., NS equations) the reader is referred to Krüger *et al.* (2017). Moreover, LBM can also solve the simplified equations derived from NS, such as the shallow water equations (Zhou, 2002).

To solve the Boltzmann equation, there are two processes that the solver needs to perform at every time step: streaming and collision. In the streaming step, each particle moves to the nearest node in the direction of its velocity. In the collision step, the particles arriving at a node, interact and change their velocity directions according to scattering rules (Chen and Doolen, 1998).

There are two main disadvantages for this formulation, which otherwise is fairly straightforward to implement. The first one is that it is memory-intensive. Krüger *et al.* (2017) note that a small case with a $100 \times 100 \times 100$ lattice requires approximately $150\,\text{MB}$ of RAM. The second one is that the method is limited to small Mach numbers (i.e., velocities much smaller than the speed of sound). However, this second shortcoming does not impact significantly the applicability of the method to coastal engineering problems, since they are subsonic.

The LBM can also deal with multiphase flows using different techniques such as the R-K color gradient (Gunstensen *et al.*, 1991),

Shan-Chen (SC) (Shan and Chen, 1993), Free energy (FE) (Swift *et al.*, 1996), He-Chen-Zhang (HCZ) (He *et al.*, 1999) or even the VOF (Hirt and Nichols, 1981; Thorimbert *et al.*, 2016) approaches. All these methods are reviewed in depth in Sudhakar and Das (2020).

6.3   Numerical Models and Applications

The Lattice–Boltzmann Method is still in an early stage of development and there are not many published works at this stage, but this technique offers promising features. One of the earlier works on LBM modeling of water waves is Zhou (2002), in which the LBM is applied to solve the shallow water equations. However, as the computational resources and the numerical method advanced, the works started to focus on solving the NS equations, which allow simulating all wave transformation processes.

Zhao *et al.* (2013) developed a 2D model capable of simulating wave propagation, and validated it with solitary and regular waves propagating over a horizontal bottom and regular waves over a submerged bar (Beji and Battjes, 1994) with excellent results. Thorimbert *et al.* (2016) simulated an Oscillating Water Column (OWC) wave-energy converter system, obtaining fair comparison with experimental results.

Most recently, Badarch *et al.* (2020) developed a single-phase solver that can simulate coastal structures. The solid portion of the structures is introduced with an IBM and the porous materials using an additional friction force. Application cases include calculating infiltration on a beach due to run-up and the hydrodynamic forces on a breakwater, which are successfully compared against Goda's formulation.

Liu *et al.* (2021) developed a 3D LBM solver and tested it against well-known benchmarks that have been widely used to validate other CFD models. The solver they developed performs as accurately as established CFD models, but it presents a more efficient scaling than CFD methods when using a large number of parallel processes for the simulations.

Finally, Hu *et al.* (2021) portrays how LBM solvers can also have additional physics coupled. In this case, Hu *et al.* (2021) simulates sediment transport in an open channel in a turbulent regime. The flow is solved with the LBM, whereas the motion and interaction between individual sediment particles are solved with a Discrete Element Method (DEM). The coupling between the particles and the fluid is performed via the Immersed Boundary Method, which has been described before. The initial validation presents a remarkable agreement with previous works, and the final simulation shows particle movements and very detailed turbulent structures near the bed.

Apart from the research models developed in the papers above, there are two free and open source frameworks that, although they have not been used yet to solve coastal applications, offer promising features. These are Palabos[1] and OpenLB.[2]

---

[1]https://palabos.unige.ch
[2]https://www.openlb.net

Part 3

Wave Modeling Couplings

The present chapter introduces the main approaches available in the literature to create a hybrid model, that links different wave models so that they can work together effectively. Another type of hybrid models exist, in which a wave model is coupled with another model capable of simulating additional physics (e.g. sediment transport, solid mechanics...). This topic has already been introduced briefly toward the end of Chapter 5, and will not be further discussed in this section.

Chapter 7

Couplings Between Wave Models

Often, when dealing with couplings between different wave modeling approaches, one of the wave models involved, which will be referred to as the "simple model" here, can be applied to a significantly larger domain or is able to simulate longer time series, as compared to the other model, which will be referred to as the "complex model". Hybrid models help to bridge limitations in modeling. For example, the complex model may not be able to simulate the complete domain without the help of the simple model due to computational limitations, or because it will take too long. Consequently, using a hybrid model is not only more efficient, but sometimes it is an enabling factor.

In general, there are two main approaches to couple numerical models, one-way and two-way couplings.

In a one-way coupling approach, the simple model is run first independently of the other. The output of the first model is stored and later used as input for the complex model, which is also run independently. This approach is the simplest option for coupling, as it does not involve modifying the models, only converting the output file from the first model to a suitable format that the second model can read. The main disadvantage of this method is that there is no feedback from the complex model to the simple model, therefore, waves evolve in the first model independently of the processes that take place in the domain of the second model. For example, if there is a reflective structure solved by the second model, the expected system of (partially) standing waves will only develop in the complex model, not being present in the simple model.

In two-way couplings both models run concurrently and communicate with each other, exchanging wave information at every time step. The main advantage of this method is that there is a two-way communication, with feedback from the complex model to the simple model. With this approach, and following the previous example, the (partially) standing waves will appear in both models, thus allowing further nonlinear wave–wave interactions to develop in the first model (if it is supported), which may modify the wave conditions incident to the second model. Two-way couplings also present several disadvantages. First, both models need to run at the same time, therefore, the computational resources required are higher. Moreover, the models need to communicate, so there is an overhead that will make both models run slower than if they were doing so independently. Also, sharing data between models can happen in two different forms, either by sharing a file or by sharing memory. Sharing a file is simple, but has a large overhead associated with writing and reading data from a hard disk. Sharing memory is much faster but much more complex, as it often requires modifying the code of the solvers so that they can ingest the data from the other model.

There are two main challenges when coupling models. The most complex issue is managing the compatibility between two sets of different equations that have been derived based on different (and even incompatible) assumptions. In some cases, this may include transforming depth-averaged data (single velocity along the water column) to a depth-varying velocity profile and the other way around. Several options to deal with this process will be discussed in the following pages. The second challenge is the compatibility of time steps between models, since it is highly likely that the complex model will require significantly smaller time steps to run. There are several options to solve this, including running each model with its own time step and interpolating linearly in time as required, or constraining both models to the smallest time step between both. The first option has the disadvantage that the feedback from the slowest model (i.e., the smallest time step) to the fastest model (i.e., the largest time step) does not occur every time step, but the simulation will run faster than selecting the second option.

Couplings present unique challenges and can be implemented in different ways, regardless of being of one- or two-way type. Spatially, the connection between the models can be limited to one side, where the domain of the complex model is an extension of the domain of the simple model. Alternatively, the domain from the complex model can be completely contained within the domain of the simple model, with connections at multiple areas.

Regardless of the number of connections, couplings can be implemented as boundary conditions or as relaxation zones. Relaxation zones exchange information in a region of the numerical domain, in which the data from one model is progressively relaxed to fit the other. This implies that the flow behavior inside the relaxation zone may not be entirely physical because its values are a blend between those resulting from two different sets of equations. As a result, the domain needs to be extended, so that the relaxation zone is far enough from the area of interest. This fact may increase the computational cost of running the case significantly, as relaxation zones can be on the order of magnitude of the wavelength. Moreover, this approach may involve duplicating a region and sharing a significant amount of data between models, since the relaxation zone may comprise a large number of cells.

Alternatively, boundary conditions exchange the data at the boundary of the numerical domain, therefore, the exchange of information is minimized and can occur without a large overhead compared to relaxation zones. Consequently, the domain does not need to be extended, minimizing the computational cost. Nevertheless, the boundary condition needs to transform the input data to be compatible with the new set of equations, so that the simulation results are physical, which is a complex process.

7.1 Coupling between phase-averaged and phase-resolving models

The first type of coupling to review is between phase-averaged and phase-resolving models. Generally, the phase-averaged approach allows simulating very large domains, much larger than phase-resolving models normally can, since the latter provide more detailed wave information. The main obstacle behind this coupling is that phase-averaged models resolve the equations in the frequency domain, while phase-resolving models do so in the time domain. In other words, phase-averaged models track the evolution of the energy level of each wave component, whereas phase-resolving models track individual waves. Such a different starting point makes developing a direct coupling impossible without additional assumptions.

Phase-averaged models provide information about the wave spectrum, which represents an infinite number of combinations of time series of free surface elevation. Generally, picking a random phase for each of the individual components of the discretized spectrum is the most straightforward way to obtain one of such time series of free surface elevation and pass the wave characteristics from a phase-averaged to a phase-resolving model. This technique is used in the vast majority of the works available in the literature, some of which are described in the next section. However, assuming that all the phases are random may not necessarily be true, especially in areas such as the surf zone (McCabe and Stansby, 2010).

One of the challenges associated with testing a single realization of a time series is that, since it has been generated in a random way, the results will also be random. Consequently, two different realizations may show very different instantaneous results. However, as usual when dealing with irregular sea states, it is the statistics that matter. As a result, since all time series derive from the same spectrum, if the simulation is run for a long-enough duration (e.g., hundreds or thousands of waves), all the simulations should produce very similar (i.e., statistically equivalent) results.

In order to transfer gravity wave information from a phase-resolving to a phase-averaged model, the time series of free surface elevation needs to be transformed to a spectrum using a Fast Fourier Transform (FFT). However, the FFT should operate in a long-enough time series to provide sufficient frequency resolution, which means that the spectrum obtained as output is not really instantaneous, as it is also built from the past values. This might pose a limitation in cases in which wave conditions change fast. Nevertheless, it is not common to perform a coupling dealing with gravity waves in this direction. Alternatively, if only long waves or tides need to be transferred, this can be done by simply updating the water depth at the relevant nodes of the phase-averaged model.

7.1.1   *Examples of couplings*

Cheung *et al.* (2003) coupled three different models (one-way) to simulate the coastal flooding resulting from storm surges and waves produced by tropical cyclones. At a first stage, the spectral model WAM (WAMDI Team, 1988) is used to calculate storm surge and wave conditions at an oceanic scale forced by the tropical cyclone. The output from WAM is used as input in the spectral model SWAN to propagate the waves nearshore and provide an estimate for wave setup. Finally, the resulting wave spectrum and storm surge data from the previous models are used along with the tide information to force the COULWAVE Boussinesq model, which simulates surf zone hydrodynamics and run-up at the coast. The model results were found to be in good accordance with real events recorded in Hawaii.

Oki and Sakai (2009) coupled a spectral model and a Boussinesq model in a 2DH domain. The one-way coupling takes place with relaxation zones at two of the sides, while the other two sides are sponge layers which absorb the waves. Several interpolation operations are required to convert the spectrum to the time series. Initially, the number of components of the input spectrum (from the spectral model) is increased using third-order spline interpolation. Then, the spectra at different locations on the spectral model grid, which has large elements, are interpolated linearly to the smaller grid of the Boussinesq model. Finally, the time series is obtained with random phases. Simple simulations with long-crested regular and irregular waves, as well as simulations in which refraction and diffraction are important, produce a good agreement with physical experiment data.

McCabe and Stansby (2010) evaluated the performance of the spectral model SWAN, a one-way coupling between SWAN and an NLSW model and a Boussinesq model for waves propagating, breaking and running up and down a slope in 1D. Comparisons of the numerical modeling data were made with the experiments of Stive (1985).

The coupling between SWAN and the NLSW model was done selecting the same frequency bands and setting the wave phase to a random value. Four different coupling locations were tested, "starting from where the reduction in wave heights starts to become apparent," since "when the coupling location is too far offshore, too much wave energy is lost too soon" (McCabe and Stansby, 2010).

The main conclusion of McCabe and Stansby (2010) highlight that the coupled model performed better than running the whole case using the Boussinesq model for the particular conditions tested in their work. However, the search for best coupling location was inconclusive, as a coupling closer to the shore produced better wave height results and a coupling further offshore produced better run-up results.

Chen *et al.* (2018) coupled SWAN and the hydrostatic hydrodynamic model FVCOM to study the interaction of waves with tides and currents. This work provides a complete information about the coupling procedure, giving full details on their implementation and including a comprehensive literature review on previous works which developed similar couplings. The two-way coupling involves transferring the free surface elevation and depth-averaged velocities (i.e., currents) from FVCOM to SWAN, and the significant wave height, mean wave length, mean wave direction and fraction of breaking waves from SWAN to FVCOM, with which the latter calculates the radiation stress (which in turn drives the currents). From the two approaches tested, running both models simultaneously was found advantageous in terms of efficiency and the adequate comparisons with experimental data.

7.2   Coupling between phase-resolving models

The procedures to couple two phase-resolving models operate in the time domain only, which is a significant advantage. However, they need to ensure the fields that are transferred are consistent with the assumptions of each model. For example, velocity and pressure profiles along the water column in CFD models can have an arbitrary shape (as required by complex hydrodynamics), and the free surface can overturn if plunging wave breaking occurs, and contain detached droplets and entrained air bubbles. If a CFD model was going to be coupled with a nonlinear shallow water (NLSW) equation model, the free surface would need to be adapted to eliminate the complex features and have a single point at all locations. Moreover, the horizontal component of velocity would need to be depth-averaged to accommodate the formulation of the NLSW model, and the vertical velocity component would then be already constrained by the NLSW continuity equation, regardless of the value in the CFD model. Further challenges would be linked to achieving a time step compatibility between both models and developing the interface, as mentioned at the start of this section.

In the case of one-way couplings, there is an additional challenge. Since no information is transferred back from the second model to the first model, "one must use a damping zone near the interface between two domains to remove the reflection and radiation waves caused by the structures" (Zhang, 2018) so as to maintain consistent wave conditions at the interface.

Such challenges are not limited to coupling these two modeling approaches, but any pair from those included in Chapter 2 will require similar procedures.

Although in principle more than two models can be coupled simultaneously, the increasing complexities that it involves often make it impractical. As mentioned before, most often a hybrid model between phase-resolving models will include a "simple" model (e.g., NLSW, Boussinesq, FNPF...) that is able to cover a large area, and a "complex" model (CFD), which will be able to solve a small area of interest with high detail, even for very complex hydrodynamic conditions. The hybrid model is advantageous, since simulating the complete domain covered by the simple model with the CFD model would be too computationally expensive.

Depending on the situation, a two-way coupling may be required, whereas in some cases a one-way coupling may suffice, as described in the following examples.

7.2.1 *Examples of couplings*

The model of Guignard *et al.* (1999) is probably one of the first examples of a hybrid model applied in coastal engineering. They coupled an FNPF model with a CFD model in a one-way fashion to develop a 2D numerical wave tank. The goal of the paper was to study wave breaking. Therefore, the waves are generated and propagated in the FNPF model, and wave breaking takes place in the CFD model. Later, in Biausser *et al.* (2003), the model was extended to three dimensions.

Sitanggang *et al.* (2007) developed a hybrid model with a two-way coupling between a Boussinesq model and a CFD model. In this hybrid model, "one model serves as the boundary condition for the other," by setting the velocities and free surface elevation at four overlapping cells/columns of cells in the grid. The Boussinesq velocities and free surface elevation are straightforward to convert to the CFD velocities and VOF magnitudes. The CFD velocities are fitted to a second-order polynomial, for which one of the components is the input value in the Boussinesq model, and the total free surface elevation is also transferred. During the phase in which the information is exchanged, convergence is sought before progressing. Tests in this paper include solitary wave propagation, which is able to travel seamlessly across the interface between both models, and a fully reflective wall with regular waves in which a standing wave system develops and extends over both models successfully. The ultimate goal of this model is to handle wave propagation with the relatively lower cost Boussinesq model and switching to the CFD model before wave breaking, so that the more complex and turbulent flow region is solved with the CFD model. A later work (Sitanggang and Lynett, 2010) provides a full description of the implementation procedures and applications to a large-scale propagation of a tsunami wave and to the overtopping of regular and solitary waves over an impermeable coastal structure.

A similar approach was presented in Kim *et al.* (2010), coupling a Boundary Element Method (BEM), which solves potential flow, with a CFD model. The numerical models are a BEM solver developed in-house and CADMAS-SURF (Arikawa *et al.*, 2007). The rationale behind this model is almost identical to that of the previous work, let the BEM model simulate the propagation of the wave under potential flow conditions, and then have the CFD model simulate the complex interactions of waves with coastal structures. However, the implementation is based on relaxation zones, which deal with an overlapping area in which the solutions of both solvers are blended in a smooth progressive way. In Figure 5 of the paper, the authors show how this procedure is able to blend the solution of both models (which might be quite different) over the area of the interface, making the free surface match. Tests include regular and irregular waves for a fully reflective and fully absorbing (sponge layer) ends, with successful results closely matching the theory.

Thorimbert *et al.* (2019) developed a two-way coupling between the shallow water equations and RANS using a Lattice and Boltzmann Method solver via boundary conditions. The results were found satisfactory as long as the waves satisfied the shallow water conditions at the coupling location.

Zhang (2018) also developed a coupling between an FNPF solver (QALE-FEM) and a CFD model based on OpenFOAM. The coupling procedure is a relaxation zone, which blends the variables smoothly over a region in space, as in the previous example. The peculiarity is that the CFD domain is contained within the FNPF domain and the coupling is only one-way. This means that the waves and currents propagate in the FNPF model without taking the CFD domain (and the structures that it may contain) into consideration. Therefore, although both models are run concurrently, they could have been run one after the other. The CFD domain is significantly smaller than the overall FNPF domain, and does not even reach the bottom of the flume, as in the previous cases. Instead, the CFD mesh is surrounded by relaxation zones in which the coupling occurs at the same time. This design minimizes the domain of the most expensive model, saving time and computational resources up to 85%, as reported. In Li *et al.* (2018), the model was applied to validation tests including steep waves interacting with currents and with a cylinder located near the free surface in a 2DV way, obtaining an adequate performance. More recently, this hybrid solver qalefOAM was employed to study the interactions between focused waves and wave-energy converter (Wang *et al.*, 2020a). This qalefOAM model demonstrated superiority in computational efficiency over using CFD model alone in recent blind tests (Ransley *et al.*, 2019, 2021).

Higuera *et al.* (2018a) developed a one-way coupling between Lagrangian, a non-hydrostatic wave model and the CFD solver olaFlow (Higuera, 2017) (based on OpenFOAM), via a boundary condition. The advantage of the Lagrangian model is that it produced very accurate wave kinematics using a moving wavemaker at a lower cost than the CFD model. In this case, the Lagrangian model was run first, generating a set of focused waves and storing the time series of relevant variables (free surface elevation, velocities and pressures) at the location of the interface to be inputted later in the CFD model. The CFD domain was significantly smaller than the Lagrangian one and contained a fixed FPSO (Floating Production Storage and Offloading) structure with which the steep waves interacted. The stored information was then interpolated linearly in space and time to be used to generate the waves in the CFD model. Active wave absorption ensured that any reflections did not re-reflect back into the domain. The main advantage of this method was that the coupling via a boundary condition does not increase the computational cost significantly, therefore, it was preferred to relaxation zones. The results of the paper were part of a blind-testing initiative, and a comparison with all the contestants is included in Ransley *et al.* (2019). The same hybrid model was later extended in Higuera *et al.* (2021) as a part of another blind modeling contest to include a combination of a boundary condition and relaxation zones, which presents advantages in terms of the efficiency of wave absorption.

Chapter 8

Executive Summary

In this book we have reviewed all the numerical modeling approaches to simulate coastal and offshore hydrodynamics. The review has been divided into phase-averaged models, simulating the evolution of the energy for each wave frequency component, and phase-resolving models, in which the time evolution of wave fields is reproduced. Generally, phase-averaged models can be applied at a large scale and are useful to evaluate how waves are generated by wind and how they propagate toward the coast. Phase-resolving models can be applied to a wide range of scales, often inversely proportional to the level of complexity of the underlying equations. This is because more sophisticated equations, requiring high computational resources, are able to reproduce complex processes more accurately, which often require finer grids.

A high-level overview of each of the methods is summarized as follows:

- **Spectral modeling** can be used to simulate the generation, propagation and dissipation of wind waves in a phase-averaged manner on a global or regional scale. However, only the spectral energy density is provided in space and time. The time history of surface elevation cannot be obtained directly due to the lack of the phase information.

- **Mild-slope equation modeling** can deal with weakly nonlinear waves, rapidly varying bottom and various types of energy dissipation sources. These models are popular to describe the wave propagation in extended areas where the combined effects of refraction and diffraction are important.

- **Nonlinear shallow water modeling** is an efficient depth-averaged approach, widely used to simulate waves in the shallow water regime (i.e., non-dispersive long waves) such as tsunamis, storm surges and overland flows. NLSW modeling is equivalent to HWM, except that it cannot resolve the stratification of the velocity profile.

- **Boussinesq modeling** is an efficient phase-resolving approach suitable for nonlinear and dispersive waves propagating from deep water to the coast. Boussinesq models can simulate nearshore wave phenomena, including reflection, refraction, diffraction and shoaling. The applicable range of most Boussinesq models is limited to $kd < \pi$, e.g., the fully nonlinear and weakly dispersive Boussinesq model FUNWAVE, and it is also restricted to potential or weak horizontal vorticity assumptions, which makes it not applicable for simulating waves interacting with vertically sheared currents.

- **Green–Naghdi modeling** is not restricted by the potential flow assumption, and it is capable of simulating not only waves alone but also waves interacting with vertically-sheared currents. The applicable range of the model in terms of wave properties and vertical profile of the current strongly depends on the underlying polynomial assumptions on the velocity profiles. Models of higher order approximations are more capable, but less efficient. Yang and Liu (2020) derived a new set of models that is superior to the Green–Naghdi models in terms of various wave properties, which has been further extended to take vertically arbitrarily sheared currents into consideration (Yang and Liu, 2022).

- **Higher-order spectral modeling** is an efficient approach to simulate ocean waves in variable water depth on a spatial-temporal scale of $\sim 10^4$ peak wavelengths in space and $\sim 10^3$ peak wave periods in time. HOS applications are limited to irrotational and inviscid flow due to the potential theory assumptions.

- **Fully nonlinear potential flow modeling** is relatively less efficient than the HOS modeling and also limited to modeling irrotational and inviscid flow. Nevertheless, it can handle complex surface-piercing fixed/floating structures.

- **Hydrostatic wave modeling** is widely used for modeling the ocean processes with stratification at global and regional scales.

This type of modeling cannot be applied to situations when the shallow water wave condition breaks down, i.e., when wavelength is comparable to water depth.

- **Non-hydrostatic wave modeling** is a convenient method to simulate nonlinear waves with high accuracy over large domains and all relative water depths. While the non-hydrostatic assumption is quite general, NHWM simulations are much less detailed than CFD. Generally, additional equations can be coupled with the hydrodynamics to solve sediment and pollutant transport problems.

- **Eulerian CFD modeling** is the most comprehensive method used to simulate coastal processes in detail, including wave breaking. This approach is widely used in research and consultancy but requires significant computational resources, therefore, it is limited to simulating at the local scale (i.e., few square kilometers at most). CFD models can also be coupled with other models or equations to increase the range of modeling, decrease the computational cost and to simulate additional physics (e.g., FSI, scour...).

- **Lagrangian CFD modeling** is fundamentally similar to Eulerian CFD modeling in terms of its range of applicability, but it offers more flexibility, as it is a meshless approach. This method can run faster using GPUs and produces realistic flow visualizations easily. Nevertheless, Lagrangian CFD is in an earlier stage of development and still needs to improve particular aspects, such as boundary conditions and pressure calculations.

- **Lattice–Boltzmann modeling** is a promising approach. It has a similar range of applicability and is as comprehensive as CFD but it also retains some fundamental physics and is easily and massively parallelizable. However, this method is still in a very early stage of development for coastal engineering applications.

Coupling different models can present significant advantages when solving coastal problems involving a wide range of scales or physics, such as the following:

- Overcome the limitations that a particular model may have in a certain area and extend the range of applicability of models. For example, coupling a Boussinesq model for wave propagation with a CFD model, to simulate wave breaking at the shoreline.

- Reduce the overall computational cost or gain accuracy at a reduced increase of computational cost. For example, the previous example will be less computationally expensive and run faster than simulating the whole domain in CFD.
- Solve additional physics complementary to hydrodynamics. For example, adding sediment transport/scour, liquefaction calculation or pollutant transport capabilities to a wave model.

There are two ways to implement couplings. For the one-way coupling, in which models run sequentially, it is simpler to develop but there is no feedback from the second model. On the other hand, for the two-way coupling, models run simultaneously and it is significantly more complex to develop, but there is full feedback between the models. The main challenges to implement hybrid models in coastal engineering are

- Developing effective coupling strategies and interfaces is complex from a technical perspective, especially when needing to match different equations with diverse underlying assumptions.
- The location of the coupling interface as a function of the relative water depth for the target waves needs to be assessed and defined in scientific and systematic ways.
- When coupling phase-averaged and phase-resolving models, the wave phases need to be set. Often they are selected randomly, but this may not always be suitable and needs to be assessed and defined in scientific and systematic ways.

Appendix

Numerical Models

Table A.1   Numerical models for coastal/offshore engineering.

Modeling approach	Numerical models
Spectral models	SWAN, TOMAWAC, CREST, WWMIII, WAM, Wavewatch III
Mild-Slope Equation (MSE)	MIKE21-EMS, MIKE21-PMS, REFDIF
Nonlinear Shallow Water equation (NLSW)	COMCOT, SCHISM, XBeach
Boussinesq equation	BOSZ, Celeris, COULWAVE, FUNWAVE, MIKE21-BOUSS
Green–Naghdi equations	GNvorti_1D, GN2D, UHAINA
High-Order Spectral (HOS) method	HOS-Ocean, ESBI
Fully Nonlinear Potential Flow (FNPF) method	OceanWave3D (FDM), BIEM, FEM, QALE-FEM, SEM
Hydrostatic wave modeling (HWM)	ADCIRC, CROCO, Delft3D, Delft3D FM, FESOM, FVCOM, HYCOM, ICON-O, MICOM, MIKE3, MITgcm, MOM6, MPAS-Ocean, NEMO, POM, PSOM, ROMS, SCHISM, SLIM, SUNTANS, UNTRIM
Non-hydrostatic wave modeling (NHWM)	Delft3D, FVCOM-NH, MIKE3, NHWAVE, SWASH, SUNTANS, XBeach-NH
Computational Fluid Dynamics (CFD) modeling	Basilisk, CADMAS-SURF, CD-Adapco, COBRAS, ComFLOW, DualSPHysics, Flow3D, Fluent, Fluidity, Gerris, GPUSPH, IH2VOF, IH3VOF, ISPH, OpenFOAM, Proteus, Reef3D, ReFRESCO, Truchas, VOFbreak
Lattice–Boltzmann method (LBM) modeling	FS-IB-LBM, OpenLB, Palabos

References

Abadie, S., Harris, J., Grilli, S., and Fabre, R. (2012). Numerical modeling of tsunami waves generated by the flank collapse of the Cumbre Vieja Volcano (La Palma, Canary Islands): Tsunami source and near field effects. *Journal of Geophysical Research: Oceans* **117**, C5, doi:https://doi.org/10.1029/2011JC007646.

Abdalla, S. and Cavaleri, L. (2002). Effect of wind variability and variable air density on wave modeling, *Journal of Geophysical Research: Oceans* **107**, C7, p. 17–1, doi:https://doi.org/10.1029/2000JC000639.

Abohadima, S. and Isobe, M. (1999). Linear and nonlinear wave diffraction using the nonlinear time dependent mild slope equations, *Coastal Engineering* **37**, 2, pp. 175–192, doi:https://doi.org/10.1016/S0378-3839(99)00020-4.

Afkhami, S., Zaleski, S., and Bussmann, M. (2009). A mesh-dependent model for applying dynamic contact angles to VOF simulations, *Journal of Computational Physics* **228**, 15, pp. 5370–5389, doi:https://doi.org/10.1016/j.jcp.2009.04.027.

Agnon, Y., Madsen, P. A., and Schäffer, H. A. (1999). A new approach to high-order Boussinesq models, *Journal of Fluid Mechanics* **399**, pp. 319–333, doi:https://doi.org/10.1017/S0022112099006394.

Al Saady, M. (2014). Numerical study of regular and irregular wave interaction with vertical breakwaters, Master's thesis, Delft University of Technology, Delft.

Altomare, C., Crespo, A. J. C., Rogers, B. D., Dominguez, J. M., Gironella, X., and Gómez-Gesteira, M. (2014). Numerical modelling of armour block sea breakwater with smoothed particle hydrodynamics, *Computers & Structures* **130**, pp. 34–45, doi:https://doi.org/10.1016/j.compstruc.2013.10.011.

Alves, J. H. and Banner, M. L. (2003). Performance of a saturation-based dissipation source term for wind wave spectral modelling, *Journal of Physical Oceanography* **33**, pp. 1274–1298, doi:https://doi.org/10.1175/1520-0485(2003)033⟨1274:POASDS⟩2.0.CO;2.

Amsden, A. A. and Harlow, F. H. (1970). A simplified MAC technique for incompressible fluid flow calculations, *Journal of Computational Physics* **6**, 2, pp. 322–325, doi:https://doi.org/10.1016/0021-9991(70)90029-X.

Ardhuin, F. (2001). Swell across the continental shelf, Tech. rep., Naval Postgraduate School, Monterrey, CA.

Ardhuin, F. and Herbers, T. H. C. (2002). Bragg scattering of random surface gravity waves by irregular seabed topography, *Journal of Fluid Mechanics* **451**, pp. 1–33, doi:https://doi.org/10.1017/S0022112001006218.

Ardhuin, F., Herbers, T. H. C., and O'Reilly, W. C. (2001). A hybrid Eulerian–Lagrangian model for spectral wave evolution with application to bottom friction on the continental shelf, *Journal of Physical Oceanography* **31**, 6, pp. 1498–1516, doi:https://doi.org/10.1175/1520-0485(2001)031⟨1498:AHELMF⟩2.0.CO;2.

Ardhuin, F. and Magne, R. (2007). Scattering of surface gravity waves by bottom topography with a current, *Journal of Fluid Mechanics* **576**, pp. 235–264, doi:https://doi.org/10.1017/S0022112006004484.

Ardhuin, F., Rawat, A., and Aucan, J. (2014). A numerical model for free infragravity waves: Definition and validation at regional and global scales, *Ocean Modelling* **77**, pp. 20–32, doi:https://doi.org/10.1016/j.ocemod.2014.02.006.

Arikawa, T., Yamano, T., and Akiyama, M. (2007). Advanced deformation method for breaking waves by using CADMAS-SURF/3D, in *Proceedings of Coastal Engineering, JSCE*, Vol. 54 (Japan Society of Civil Engineers), pp. 71–75, doi:https://doi.org/10.2208/proce1989.54.71.

Ashgriz, N. and Poo, J. Y. (1991). FLAIR: Flux line-segment model for advection and interface reconstruction, *Journal of Computational Physics* **93**, 2, pp. 449–468, doi:https://doi.org/10.1016/0021-9991(91)90194-P.

Babanin, A. and Young, I. (2005). Two-phase behaviour of the spectral dissipation of wind waves, in *Proceedings of the 5th International Symposium Ocean Wave Measurement and Analysis*, Madrid, June 2005.

Badarch, A., Fenton, J. D., and Hosoyamada, T. (2020). Application of free-surface immersed-boundary Lattice Boltzmann method to waves acting on coastal structures, *Journal of Hydraulic Engineering* **146**, 2, doi:https://doi.org/10.1061/(ASCE)HY.1943-7900.0001679.

Baker, G. R., Meiron, D. I., and Orszag, S. A. (1982). Generalized vortex methods for free-surface flow problems, *Journal of Fluid Mechanics* **123**, pp. 477–501, doi:https://doi.org/10.1017/S0022112082003164.

Banner, M. L. and Melville, W. K. (1976). On the separation of air flow over water waves, *Journal of Fluid Mechanics* **77**, 4, pp. 825–842, doi:https://doi.org/10.1017/S0022112076002905.

Barthélemy, E. (2004). Nonlinear shallow water theories for coastal waves, *Surveys in Geophysics* **25**, pp. 315–337, doi:https://doi.org/10.1007/s10712-003-1281-7.

Bateman, W. J., Swan, C., and Taylor, P. H. (2001). On the efficient numerical simulation of directionally spread surface water waves, *Journal of Computational Physics* **174**, 1, pp. 277–305, doi:https://doi.org/10.1006/jcph.2001.6906.

Bateman, W. J., Swan, C., and Taylor, P. H. (2003). On the calculation of the water particle kinematics arising in a directionally spread wavefield, *Journal of Computational Physics* **186**, 1, pp. 70–92, doi:https://doi.org/10.1016/S0021-9991(03)00012-3.

Battjes, J. A. and Janssen, J. P. F. M. (1978). Energy loss and set-up due to breaking of random waves, in *Proceedings of Coastal Engineering Conference 1978*, pp. 569–587, doi:https://doi.org/10.1061/9780872621909.034.

Baykal, C., Sumer, B. M., Fuhrman, D. R., Jacobsen, N. G., and Fredsøe, J. (2017). Numerical simulation of scour and backfilling processes around a circular pile in waves, *Coastal Engineering* **122**, pp. 87–107, doi:https://doi.org/10.1016/j.coastaleng.2017.01.004.

Beji, S. and Battjes, J. A. (1993). Experimental investigation of wave propagation over a bar, *Coastal Engineering* **19**, 1–2, pp. 151–162, doi:https://doi.org/10.1016/0378-3839(93)90022-Z.

Beji, S. and Battjes, J. A. (1994). Numerical simulation of nonlinear wave propagation over a bar, *Coastal Engineering* **23**, 1–2, pp. 1–16, doi:https://doi.org/10.1016/0378-3839(94)90012-4.

Benoit, M., Marcos, F., and Becq, F. (1997). Development of a third generation shallow-water wave model with unstructured spatial meshing, in *Proceedings of Coastal Engineering Conference 1996*, pp. 465–478, doi:https://doi.org/10.1061/9780784402429.037.

Berkhoff, J. C., Booy, N., and Radder, A. (1982). Verification of numerical wave propagation models for simple harmonic linear water waves, *Coastal Engineering* **6**, pp. 255–279, doi:https://doi.org/10.1016/0378-3839(82)90022-9.

Berkhoff, J. C. W. (1972). Computation of combined refraction-diffraction, in *Proceedings of Coastal Engineering Conference 1972*, pp. 471–490, doi:https://doi.org/10.9753/icce.v13.23.

Biausser, B., Grilli, S., and Fraunié, P. (2003). Numerical simulations of three-dimensional wave breaking by coupling of a VOF method and a boundary element method, in *Proceedings of the 13th International Offshore and Polar Engineering Conference*.

Bihs, H., Chella, M. A., Kamath, A., and Arntsen, Ø. A. (2017). Numerical investigation of focused waves and their interaction with a vertical cylinder using REEF3D, *Journal of Offshore Mechanics and Arctic Engineering* **139**, 4, doi:https://doi.org/10.1115/1.4036206.

Bihs, H., Kamath, A., Alagan Chella, M., and Arntsen, Ø. A. (2019). Extreme wave generation, breaking, and impact simulations using wave packets in REEF3D, *Journal of Offshore Mechanics and Arctic Engineering* **141**, 4, doi:https://doi.org/10.1115/1.4042178.

Bijvelds, M. D. J. P. (2001). Numerical modelling of estuarine flow over steep topography, PhD thesis, TU Delft, NL.

Bingham, H. B. and Zhang, H. (2007). On the accuracy of finite-difference solutions for nonlinear water waves, *Journal of Engineering Mathematics* **58**, pp. 211–228, doi:https://doi.org/10.1007/s10665-006-9108-4.

Biot, M. A. (1941). General theory of three-dimensional consolidation, *Journal of Applied Physics* **12**, 2, pp. 155–164, doi:https://doi.org/10.1063/1.1712886.

Bistafa, S. R. (2018). On the development of the Navier-Stokes equation by Navier, *Revista Brasileira de Ensino de Física* **40**, 2, doi:https://doi.org/10.1590/1806-9126-RBEF-2017-0239.

Bleck, R. and Smith, L. T. (1990). A wind-driven isopycnic coordinate model of the north and equatorial Atlantic Ocean: 1. Model development and supporting experiments, *Journal of Geophysical Research: Oceans* **95**, C3, pp. 3273–3285, doi:https://doi.org/10.1029/JC095iC03p03273.

Blumberg, A. F. and Mellor, G. L. (1987). *A Description of a Three-Dimensional Coastal Ocean Circulation Model* (American Geophysical Union (AGU)), pp. 1–16, doi:https://doi.org/10.1029/CO004p0001.

Bokaris, J. and Anastasiou, K. (2003). Solution of the hyperbolic mild-slope equation using the finite volume method, *International Journal for Numerical Methods in Fluids* **41**, 3, pp. 225–250, doi:https://doi.org/10.1002/fld.429.

Bonnefoy, F., Ducrozet, G., Touzé, D. L., and Ferrant, P. (2010). Time domain simulation of nonlinear water waves using spectral methods, in Quingwei Ma (ed.), *Advances in Numerical Simulation of Nonlinear Water Waves* (World Scientific, Singapore), pp. 129–164, doi:https://doi.org/10.1142/9789812836502_0004.

Bonnefoy, F., Touze, L., Ferrant, P., *et al.* (2004). Generation of fully-nonlinear prescribed wave fields using a high-order spectral model, in *Proceedings of the 14th International Offshore and Polar Engineering Conference*.

Bonneton, P., Chazel, F., Lannes, D., Marche, F., and Tissier, M. (2011). A splitting approach for the fully nonlinear and weakly dispersive Green–Naghdi model, *Journal of Computational Physics* **230**, 4, pp. 1479–1498, doi:https://doi.org/10.1016/j.jcp.2010.11.015.

Boo, S. Y., Kim, C. H., and Kim, M. H. (1994). A numerical wave tank for nonlinear irregular waves by 3-D higher order boundary element method, *International Journal of Offshore and Polar Engineering* **4**, 04, doi:https://doi.org/10.1029/98JC02622.

Booij, N. (1981). Gravity waves on water with non-uniform depth and current, PhD thesis, Delft University of Technology, Department of Civil Engineering, NL.

Booij, N. (1983). A note on the accuracy of the mild-slope equation, *Coastal Engineering* **7**, 3, pp. 191–203, doi:https://doi.org/10.1016/0378-3839(83)90017-0.

Booij, N. R. R. C., Ris, R. C., and Holthuijsen, L. H. (1999). A third-generation wave model for coastal regions: 1. Model description and validation, *Journal of Geophysical Research: Oceans* **104**, C4, pp. 7649–7666, doi:https://doi.org/10.1029/98JC02622.

Boussinesq, J. (1872). Théorie des ondes et des remous qui se propagent le long d'un canal rectangulaire horizontal, en communiquant au liquide contenu dans ce canal des vitesses sensiblement pareilles de la surface au fond (in French – Theory of waves and eddies that propagate along a horizontal rectangular channel, interacting with the liquid contained in this channel with similar speeds from the surface to the bottom), *Journal de Mathématiques Pures et Appliquées – Journal of Pure and Applied Mathematics* **17**, pp. 55–108, doi:https://eudml.org/doc/234248.

Brackbill, J. U., Kothe, D. B., and Zemach, C. (1992). A continuum method for modeling surface tension, *Journal of Computational Physics* **100**, 2, pp. 335–354, doi:https://doi.org/10.1016/0021-9991(92)90240-Y.

Brandini, C. and Grilli, S. T. (2001a). Modeling of freak wave generation in a 3D-NWT, in *Proceedings of the 11th International Offshore and Polar Engineering Conference*.

Brandini, C. and Grilli, S. T. (2001b). Three-dimensional wave focusing in fully nonlinear wave models, in *Proceedings of of the 4th International Symposium on Ocean Wave Measurement and Analysis*, pp. 1102–1111, doi: https://doi.org/10.1061/40604(273)112.

Breivik, Ø., Mogensen, K., Bidlot, J.-R., Balmaseda, M. A., and Janssen, P. A. E. M. (2015). Surface wave effects in the NEMO ocean model: Forced and coupled experiments, *Journal of Geophysical Research: Oceans* **120**, 4, pp. 2973–2992, doi:https://doi.org/10.1002/2014JC010565.

Brinkkemper, J. A., Torres-Freyermuth, A., Mendoza, E. T., Salles, P., and Ruessink, B. G. (2013). Parameterization of wave run-up on beaches in Yucatan, Mexico: A numerical study, in *Proceedings of Coastal Dynamics 2013*, pp. 225–233.

Brocchini, M., Bernetti, R., Mancinelli, A., and Albertini, G. (2001). An efficient solver for nearshore flows based on the WAF method, *Coastal Engineering* **43**, 2, pp. 105–129, doi:https://doi.org/10.1016/S0378-3839(01)00009-6.

Bryan, K. (1969). A numerical method for the study of the circulation of the world ocean, *Journal of Computational Physics* **4**, 3, pp. 347–376, doi:https://doi.org/10.1016/0021-9991(69)90004-7.

Burcharth, H. and Andersen, O. (1995). On the one-dimensional steady and unsteady porous flow equations, *Coastal Engineering* **24**, 3–4, pp. 233–257, doi:https://doi.org/10.1016/0378-3839(94)00025-S.

Caetano, E. and Innocentini, V. (2003). An improved second generation wave model, *Revista Brasileira de Oceanografia* **51**, pp. 1–21, doi:https://doi.org/10.1590/S1413-77392003000100002.

Cai, Y., Agrawal, A., Qu, K., and Tang, H. (2018). Numerical investigation of connection forces of a coastal bridge deck impacted by solitary waves, *Journal of Bridge Engineering* **23**, 1, doi:https://doi.org/10.1061/(ASCE)BE.1943-5592.0001135.

Cao, H., Feng, W., Hu, Z., Suzuki, T., and Stive, M. J. (2015). Numerical modeling of vegetation-induced dissipation using an extended mild-slope equation, *Ocean Engineering* **110**, pp. 258–269, doi:https://doi.org/10.1016/j.oceaneng.2015.09.057.

Castro, A. and Lannes, D. (2014). Fully nonlinear long-wave models in the presence of vorticity, *Journal of Fluid Mechanics* **759**, pp. 642–675, doi: https://doi.org/10.1017/jfm.2014.593.

Castro, M. J. and Semplice, M. (2019). Third-and fourth-order well-balanced schemes for the shallow water equations based on the CWENO reconstruction, *International Journal for Numerical Methods in Fluids* **89**, 8, pp. 304–325, doi:https://doi.org/10.1002/fld.4700.

Casulli, V. and Walters, R. A. (2000). An unstructured grid, three-dimensional model based on the shallow water equations, *International Journal for Numerical Methods in Fluids* **32**, 3, pp. 331–348, doi:https://doi.org/10.1002/(SICI)1097-0363(20000215)32:3⟨331::AID-FLD941⟩3.0.CO;2-C.

Casulli, V. and Zanolli, P. (2002). Semi-implicit numerical modeling of nonhydrostatic free-surface flows for environmental problems, *Mathematical and Computer Modelling* **36**, 9–10, pp. 1131–1149, doi:https://doi.org/10.1016/S0895-7177(02)00264-9.

Cavaleri, L., Abdalla, S., Benetazzo, A., Bertotti, L., Bidlot, J.-R., Breivik, Ø., Carniel, S., Jensen, R., Portilla-Yandun, J., Rogers, W., Roland, A., Sanchez-Arila, A., Smith, J.M., Staneva, J., Toledo, Y., Van Vledder, G.Ph., and Van der Westhuysen, A. J., (2018). Wave modelling in coastal and inner seas, *Progress in Oceanography* **167**, pp. 164–233, doi: https://doi.org/10.1016/j.pocean.2018.03.010.

Cavaleri, L., Alves, J.-H., Ardhuin, F., Babanin, A., Banner, M., Belibassakis, K., Benoit, M., Donelan, M., Groeneweg, J., Herbers, T., Hwang, P., Janssen, P.A.E.M., Janssen, T., Lavrenov, I.V., Magne, R., Monbaliu, J., Onorato, M., Polnikov, V., Resio, D., Rogers, W. E., Sheremet, A., Mckee Smith, J.,Tolman, H.L., van Vledder, G., Wolf, J., and Young, I. (The WISE Group) (2007). Wave modelling–the state of the art, *Progress in Oceanography* **75**, 4, pp. 603–674, doi:https://doi.org/10.1016/j.pocean.2007.05.005.

Cea, L. and Vázquez-Cendón, M. E. (2007). Depth averaged turbulence models and source terms, in P. G. Navarro and E. Playán (eds.), *Numerical Modelling of Hydrodynamics for Water Resources* (CRC Press, Florida, USA), pp. 139–164, doi:https://doi.org/10.1201/9781482288513.

Celebi, M. S., Kim, M. H., and Beck, R. F. (1998). Fully nonlinear 3-D numerical wave tank simulation, *Journal of Ship Research* **42**, 1, pp. 33–45, doi:https://doi.org/10.1016/j.pocean.2018.03.010.

Chamberlain, P. G. and Porter, D. (1995). The modified mild-slope equation, *Journal of Fluid Mechanics* **291**, pp. 393–407, doi:https://doi.org/10.1017/S0022112095002758.

Chandar, D. D., Boppana, B., and Kumar, V. (2018). A comparative study of different overset grid solvers between OpenFOAM, StarCCM+ and Ansys-Fluent, in *Proceedings of the 2018 AIAA Aerospace Sciences Meeting*, p. 1564, doi:https://doi.org/10.2514/6.2018-1564.

Chandrasekera, C. N. and Cheung, K. F. (2001). Linear refraction-diffraction model for steep bathymetry, *Journal of Waterway, Port, Coastal and Ocean Engineering* **127**, 3, pp. 161–170, doi:https://doi.org/10.1061/(ASCE)0733-950X(2001)127:3(161).

Chaudhry, M. H. (2007). *Open-Channel Flow* (Springer, NY, USA).

Chazel, F., Lannes, D., and Marche, F. (2011). Numerical simulation of strongly nonlinear and dispersive waves using a Green–Naghdi model, *Journal of Scientific Computing* **48**, 1–3, pp. 105–116, doi:https://doi.org/10.1007/s10915-010-9395-9.

Chen, B. Q., Kirby, J. T., Dalrymple, R. A., and Kennedy, A. B. (2000). Boussinesq modeling of wave transformation, breaking, and runup. II: 2D, *Journal of Waterway, Port, Coastal, and Ocean Engineering* **126(1)**, pp. 48–56, doi:https://doi.org/10.1061/(ASCE)0733-950X(2000)126:1(48).

Chen, H. and Zou, Q.-P. (2019). Eulerian–Lagrangian flow-vegetation interaction model using immersed boundary method and OpenFOAM, *Advances in Water Resources* **126**, pp. 176–192, doi:https://doi.org/10.1016/j. advwatres.2019.02.006.

Chen, Q. (2006). Fully nonlinear Boussinesq-type equations for waves and currents over porous beds, *Journal of Engineering Mechanics* **132**, pp. 220–230, doi: 10.1061/(ASCE)0733-9399(2006)132:2(220).

Chen, Q., Dalrymple, R. A., Kirby, J. T., Kennedy, A. B., and Haller, M. C. (1999). Boussinesq modeling of a rip current system, *Journal of Geophysical Research* **104637**, 15, pp. 617–620, doi:https://doi.org/10.1029/ 1999JC900154.

Chen, Q., Kaihatu, J. M., and Hwang, P. A. (2004). Incorporation of wind effects into Boussinesq wave models, *Journal of Waterway, Port, Coastal and Ocean Engineering* **130**, 6, pp. 312–321, doi:https://doi.org/10.1061/ (ASCE)0733-950X(2004)130:6(312).

Chen, S. and Doolen, G. D. (1998). Lattice Boltzmann method for fluid flows, *Annual Review of Fluid Mechanics* **30**, 1, pp. 329–364, doi:https://doi.org/ 10.1146/annurev.fluid.30.1.329.

Chen, T., Zhang, Q., Wu, Y., Ji, C., Yang, J., and Liu, G. (2018). Development of a wave-current model through coupling of FVCOM and SWAN, *Ocean Engineering* **164**, pp. 443–454, doi:https://doi.org/10.1016/j.oceaneng. 2018.06.062.

Chen, W., Panchang, V., and Demirbilek, Z. (2005). On the modeling of wave–current interaction using the elliptic mild-slope wave equation, *Ocean Engineering* **32**, 17–18, pp. 2135–2164, doi:https://doi.org/10.1016/ j.oceaneng.2005.02.010.

Chen, X. (2003). A fully hydrodynamic model for three-dimensional, free-surface flows, *International Journal for Numerical Methods in Fluids* **42**, 9, pp. 929–952, doi:https://doi.org/10.1002/fld.557.

Chen, Y. and Liu, P. L.-F. (1995). Modified Boussinesq equations and associated parabolic models for water wave propagation, *Journal of Fluid Mechanics* **288**, pp. 351–381, doi:https://doi.org/10.1017/S0022112095001170.

Cheung, K. F., Phadke, A. C., Wei, Y., Rojas, R., Douyere, Y. J.-M., Martino, C. D., Houston, S. H., Liu, P. L.-F., Lynett, P. J., Dodd, N., Liao, S., and Nakazaki, E. (2003). Modeling of storm-induced coastal flooding for emergency management, *Ocean Engineering* **30**, 11, pp. 1353–1386, doi: https://doi.org/10.1016/S0029-8018(02)00133-6.

Choi, B. H., Kim, D. C., Pelinovsky, E., and Woo, S. B. (2007). Three-dimensional simulation of tsunami run-up around conical island, *Coastal Engineering* **54**, 8, pp. 618–629, doi:https://doi.org/10.1016/j.coastaleng.2007.02.001.

Choi, J., Kirby, J. T., and Yoon, S. B. (2015). Boussinesq modeling of longshore currents in the SandyDuck experiment under directional random wave conditions, *Coastal Engineering* **101**, pp. 17–34, doi:https://doi.org/10.1016/ j.coastaleng.2015.04.005.

Choi, W. (2009). Nonlinear surface waves interacting with a linear shear current, *Mathematics and Computers in Simulation* **80**, 1, pp. 29–36, doi:https:// doi.org/10.1016/j.matcom.2009.06.021.

Chun, J., Song, C. G., and Ahn, K. (2013). A simple numerical method on the partial reflection and transmission of water waves in the hyperbolic mild-slope equation, *Journal of Coastal Research* **29**, 3, pp. 717–726, doi:https://doi.org/10.2112/JCOASTRES-D-12-00007.1.

Clamond, D., Fructus, D., Grue, J., and Kristiansen, Ø. (2005). An efficient model for three-dimensional surface wave simulations. Part II: Generation and absorption, *Journal of Computational Physics* **205**, 2, pp. 686–705, doi: https://doi.org/10.1016/j.jcp.2004.11.038.

Clamond, D. and Grue, J. (2001). A fast method for fully nonlinear water-wave computations, *Journal of Fluid Mechanics* **447**, pp. 337–355, doi:https://doi.org/10.1017/S0022112001006000.

Clauss, G. F. and Steinhagen, U. (1999). Numerical simulation of nonlinear transient waves and its validation by laboratory data, in *Proceedings of the 9th International Offshore and Polar Engineering Conference*, Brest, France, pp. 368–375.

Collins, J. I. (1972). Prediction of shallow-water spectra, *Journal of Geophysical Research* **77**, 15, pp. 2693–2707, doi:https://doi.org/10.1029/JC077i015p02693.

Copeland, G. J. (1985). A practical alternative to the "mild-slope" wave equation, *Coastal Engineering* **29**, 2, pp. 125–149, doi:https://doi.org/10.1016/0378-3839(85)90002-X.

Craig, W. and Sulem, C. (1993). Numerical simulation of gravity waves, *Journal of Computational Physics* **108**, 1, pp. 73–83, doi:https://doi.org/10.1006/jcph.1993.1164.

Cui, L., Jeng, D.-S., and Liu, J. (2021). Seabed foundation stability around offshore detached breakwaters, *Applied Ocean Research* **111**, doi:https://doi.org/10.1016/j.apor.2021.102672.

Cui, L., Jiang, H. Z., Chen, X. L., Yuan, Z. J., Yu, D. T., and Wu, L. G. (2014). Numerical models for parabolic mild slope equation and nearshore current in curvilinear coordinates (in Chinese), *Journal of Basic Science and Engineering* **22**, pp. 1204–1216.

D'Alessandro, F. and Tomasicchio, G. R. (2008). The BCI criterion for the initiation of breaking process in Boussinesq-type equations wave models, *Coastal Engineering* **55**, 12, pp. 1174–1184, doi:https://doi.org/10.1016/j.coastaleng.2008.05.002.

Dalrymple, R. A., Herault, A., Bilotta, G., and Farahani, R. J. (2010). GPU-accelerated SPH model for water waves and other free surface flows, *Proceedings of the 32nd International Conference on Coastal Engineering*, Shanghai.

Dalrymple, R. A. and Kirby, J. T. (1988). Models for very wide-angle water waves and wave diffraction, *Journal of Fluid Mechanics* **192**, pp. 33–50, doi:https://doi.org/10.1017/S0022112088001776.

Dalrymple, R. A., Kirby, J. T., and Mann, D. W. (1984). Wave propagation in the vicinity of islands, in *Proceedings of Offshore Technology Conference*.

Dalrymple, R. A. and Rogers, B. D. (2006). Numerical modeling of water waves with the SPH method, *Coastal Engineering* **53**, 2–3, pp. 141–147, doi:https://doi.org/10.1016/j.coastaleng.2005.10.004.

Dalrymple, R. A., Suh, K. D., Kirby, J. T., and Chae, J. W. (1989). Models for very wide-angle water waves and wave diffraction. Part 2. Irregular bathymetry, *Journal of Fluid Mechanics* **201**, pp. 299–322, doi:https://doi.org/10.1017/S0022112089000959.

Day, S. J., Watts, P., Grilli, S. T., and Kirby, J. T. (2005). Mechanical models of the 1975 Kalapana, Hawaii earthquake and tsunami, *Marine Geology* **215**, 1–2, pp. 59–92, doi:https://doi.org/10.1016/j.margeo.2004.11.008.

de Saint-Venant, A. J.-C. B. (1871). Théorie du mouvement non-permanent des eaux, avec application aux crues des rivières et à l'introduction des marées dans leur lit (in French – Theory of the non-permanent movement of water, with application to the flooding of rivers and the propagation of tides into their beds), *Comptes Rendus de l'Académie des Sciences – Reports of the Academy of Sciences* **73**, 147–154, p. 5.

Dean, R. G. and Dalrymple, R. A. (1991). *Water Wave Mechanics for Engineers and Scientists, Advanced Series on Ocean Engineering*, Vol. 2 (World Scientific, Singapore).

del Jesus, M., Lara, J. L., and Losada, I. J. (2012). Three-dimensional interaction of waves and porous structures. Part I: Numerical model formulation, *Coastal Engineering* **64**, pp. 57–72, doi:https://doi.org/10.1016/j.coastaleng.2012.01.008.

Delestre, O., Lucas, C., Ksinant, P.-A., Darboux, F., Laguerre, C., Vo, T.-N.-T., James, F., and Cordier, S. (2013). Swashes: A compilation of shallow water analytic solutions for hydraulic and environmental studies, *International Journal for Numerical Methods in Fluids* **72**, 3, pp. 269–300.

Demirbilek, Z. and Webster, W. C. (1992). Application of the Green-Naghdi theory of fluid sheets to shallow-water wave problems. Report 1. Model development, Tech. rep., Coastal Engineering Research Center, Vicksburg, MS.

Demirbilek, Z. and Webster, W. C. (1999). The Green–Naghdi theory of fluid sheets for shallow-water waves, chap. 1, in J. B. Herbich (ed.), *Developments in Offshore Engineering* (Elsevier, Gulf Publishing Company, TX, USA), pp. 1–54, doi:https://doi.org/10.1016/b978-088415380-1/50020-0.

Dentale, F., Donnarumma, G., and Carratelli, E. P. (2014). Simulation of flow within armour blocks in a breakwater, *Journal of Coastal Research* **295**, pp. 528–536, doi:https://doi.org/10.2112/JCOASTRES-D-13-00035.1.

Deshpande, S. S., Anumolu, L., and Trujillo, M. F. (2012). Evaluating the performance of the two-phase flow solver interFoam, *Computational Science & Discovery* **5**, 1, doi:https://doi.org/10.1088/1749-4699/5/1/014016.

Devolder, B., Rauwoens, P., and Troch, P. (2017). Application of a buoyancy-modified k-omega SST turbulence model to simulate wave run-up around a monopile subjected to regular waves using OpenFOAM, *Coastal Engineering* **125**, pp. 81–94, doi:https://doi.org/10.1016/j.coastaleng.2017.04.004.

DHI (2020a). MIKE 3 Flow Model FM. Hydrodynamic and transport module, Scientific documentation, Tech. rep., DHI, Netherland, doi:https://manuals.mikepoweredbydhi.help/2020/Coast_and_SeaMIKE_3_Flow_FM_Scientific_Doc.pdf, NL.

DHI (2020b). MIKE 3 Flow Model FM, Hydrodynamic module, Non-hydrostatic formulation. Validation report, Tech. rep., DHI, Netherland, doi:https://manuals.mikepoweredbydhi.help/2020/Coast_and_Sea/MIKE_3_Flow_FM_Validation_Report.pdf, NL.

Didier, E., Teixeira, P. R. F., and Neves, M. G. (2016). A 3D numerical wave tank for coastal engineering studies, *Defect and Diffusion Forum* **372**, pp. 1–10, doi:https://doi.org/10.4028/www.scientific.net/DDF.372.1.

Dimakopoulos, A. and Higuera, P. (2021). Wave generation and absorption techniques, chap. 1, in D. M. Kelly, A. Dimakopoulos, P. Higuera (eds.), *Advanced Numerical Modelling of Wave Structure Interactions* (CRC Press, Florida, USA), doi:https://doi.org/10.1201/9781351119542.

Ding, B., Wuillaume, P.-Y., Meng, F., Babarit, A., Schubert, B., Sergiienko, N., and Cazzolato, B. (2019). Comparison of wave-body interaction modelling methods for the study of reactively controlled point absorber wave energy converter, in *Proceedings of the 34th International Workshop on Water Waves and Floating Bodies (IWWWFB)*.

Dingemans, M. W. (1997). *Water Wave Propagation over Uneven Bottoms.* (World Scientific) p. 13.

Dodd, N. (1998). Numerical model of wave run-up, overtopping, and regeneration, *Journal of Waterway, Port, Coastal, and Ocean Engineering* **124**, 2, pp. 73–81, doi:https://doi.org/10.1061/(ASCE)0733-950X(1998)124:2(73).

Dold, J. W. (1992). An efficient surface-integral algorithm applied to unsteady gravity waves, *Journal of Computational Physics* **103**, 1, pp. 90–115, doi:https://doi.org/10.1016/0021-9991(92)90327-U.

Dommermuth, D. G. and Yue, D. K. P. (1987). A high-order spectral method for the study of nonlinear gravity waves, *Journal of Fluid Mechanics* **184**, pp. 267–288, doi:https://doi.org/10.1017/S002211208700288X.

Donelan, M. A. (2001). A nonlinear dissipation function due to wave breaking, in *Proceedings of ECMWF Workshop on Ocean Wave Forecasting*, pp. 87–94.

Dong, G., Wang, G., Ma, X., and Ma, Y. (2010). Harbor resonance induced by subaerial landslide-generated impact waves, *Ocean Engineering* **37**, 10, pp. 927–934, doi:https://doi.org/10.1016/j.oceaneng.2010.03.005.

Dong, P. (1987). The computation of wave-induced circulations with wave current interaction and refined turbulence modelling, PhD thesis, Imperial College, London.

Duan, W., Zheng, K., Zhao, B., Demirbilek, Z., Ertekin, R., and Webster, W. (2016). On wave–current interaction by the Green–Naghdi equations in shallow water, *Natural Hazards* **84**, 2, pp. 567–583, doi:https://doi.org/10.1007/s11069-016-2464-0.

Duan, W. Y., Wang, Z., Zhao, B. B., Ertekin, R. C., and Yang, W. Q. (2018). Steady solution of solitary wave and linear shear current interaction, *Applied Mathematical Modelling* **60**, pp. 354–369, doi:https://doi.org/10.1016/j.apm.2018.03.032.

Ducrozet, G., Bonnefoy, F., Le Touzé, D., and Ferrant, P. (2006). Implementation and validation of nonlinear wavemaker models in a HOS numerical wave tank, *International Journal of Offshore and Polar Engineering* **16**, 03.

Ducrozet, G., Bonnefoy, F., Le Touzé, D., and Ferrant, P. (2012). A modified high-order spectral method for wavemaker modeling in a numerical wave tank, *European Journal of Mechanics-B/Fluids* **34**, pp. 19–34, doi:https://doi.org/10.1016/j.euromechflu.2012.01.017.

Ducrozet, G., Bonnefoy, F., Le Touzé, D., and Ferrant, P. (2016). HOS-ocean: Open-source solver for nonlinear waves in open ocean based on high-order spectral method, *Computer Physics Communications* **203**, pp. 245–254, doi:https://doi.org/10.1016/j.cpc.2016.02.017.

Ducrozet, G., Bonnefoy, F., Touzé, D. L., and Ferrant, P. (2007). 3-D HOS simulations of extreme waves in open seas, *Natural Hazards and Earth System Sciences* **7**, 1, pp. 109–122, doi:https://doi.org/10.5194/nhess-7-109-2007.

Eckart, C. (1952). The propagation of gravity waves from deep to shallow water, *National Bureau of Standards* **20**, pp. 165–173.

Einfeldt, B. (1988). On Godunov-type methods for gas dynamics, *SIAM Journal on Numerical Analysis* **25**, 2, pp. 294–318, doi:https://doi.org/10.1137/0725021.

Eldeberky, Y. (1996). *Nonlinear Transformation of Wave Spectra in the Nearshore Zone*, PhD thesis, Delft University of Technology, Delft, Department of Civil Engineering, NL, doi:http://resolver.tudelft.nl/uuid:707ca57d-81c3-4103-bc6e-aae1c90fce63.

Eldeberky, Y. and Battjes, J. A. (1995). Parameterization of triad interaction in wave energy model, in *Proceedings of the Coastal Dynamics Conference* Gdansk, Poland.

Enet, F. (2006). Tsunami generation by under water landslides, PhD thesis, University of Rhode Island, USA.

Engsig-Karup, A., Bingham, H., and Lindberg, O. (2009). An efficient flexible-order model for 3D nonlinear water waves, *Journal of Computational Physics* **228**, 6, pp. 2100–2118, doi:https://doi.org/10.1016/j.jcp.2008.11.028.

Engsig-Karup, A. P., Eskilsson, C., and Bigoni, D. (2016). A stabilised nodal spectral element method for fully nonlinear water waves, *Journal of Computational Physics* **318**, pp. 1–21, doi:https://doi.org/10.1016/j.jcp.2016.04.060.

Engsig-Karup, A. P., Madsen, M. G., and Glimberg, S. L. (2012). A massively parallel GPU-accelerated model for analysis of fully nonlinear free surface waves, *International Journal for Numerical Methods in Fluids* **70**, 1, pp. 20–36, doi:https://doi.org/10.1002/fld.2675.

Ertekin, R. C., Webster, W. C., Ertekin, R. C., Webster, W. C., and Wehausen, J. V. (1986). Waves caused by a moving disturbance in a shallow channel of finite width, *Journal of Fluid Mechanics* **169**, pp. 275–292, doi:https://doi.org/10.1017/S0022112086000630.

Fairley, I., Masters, I., and Karunarathna, H. (2015). The cumulative impact of tidal stream turbine arrays on sediment transport in the Pentland Firth, *Renewable Energy* **80**, pp. 755–769, doi:https://doi.org/10.1016/j.renene.2015.03.004.

Ferrant, P. (1997). Simulation of strongly nonlinear wave generation and wave-body interactions using a 3-D MEL model, in *Proceedings of the 21st ONR Symposium on Naval Hydrodynamics*, Trondheim, Norway, pp. 93–109.

Fochesato, C., Grilli, S., and Dias, F. (2007). Numerical modeling of extreme rogue waves generated by directional energy focusing, *Wave Motion* **44**, 5, pp. 395–416, doi:https://doi.org/10.1016/j.wavemoti.2007.01.003.

Forcheimer, P. (1901). Wasserbewegung durch Boden (in German – Water movement through soil), *Zeitschrift des Vereins Deutscher Ingenieure – Journal of the Association of German Engineers* **45**, pp. 1782–1788.

Fox-Kemper, B., Adcroft, A., Böning, C. W., Chassignet, E. P., Curchitser, E., Danabasoglu, G., Eden, C., England, M. H., Gerdes, R., Greatbatch, R. J., Griffies, S. M., Halberg, R. W., Hanert, E., Heimbach, P., Hewit, H. T., Hill, C. N., Komuro, Y., Legg, S., Le Sommer, J., Masina, S., Marsland, S. J., Penny, S. G., Qiao, F., Ringler T. D., Treguier, A. M., Tsujino, H., Uotila, P., and Yeager, S. G. (2019). Challenges and prospects in ocean circulation models, *Frontiers in Marine Science* **6**, p. 65, doi:https://doi.org/10.3389/fmars.2019.00065.

Francois, M. M., Cummins, S. J., Dendy, E. D., Kothe, D. B., Sicilian, J. M., and Williams, M. W. (2006). A balanced-force algorithm for continuous and sharp interfacial surface tension models within a volume tracking framework, *Journal of Computational Physics* **213**, pp. 141–173, doi:https://doi.org/10.1016/j.jcp.2005.08.004.

Fringer, O., Gerritsen, M., and Street, R. (2006). An unstructured-grid, finite-volume, nonhydrostatic, parallel coastal ocean simulator, *Ocean Modelling* **14**, 3–4, pp. 139–173, doi:https://doi.org/10.1016/j.ocemod.2006.03.006.

Fructus, D., Clamond, D., Grue, J., and Kristiansen, Ø. (2005). An efficient model for three-dimensional surface wave simulations: Part I: Free space problems, *Journal of Computational Physics* **205**, 2, pp. 665–685, doi:https://doi.org/10.1016/j.jcp.2004.11.027.

Fructus, D. and Grue, J. (2007). An explicit method for the nonlinear interaction between water waves and variable and moving bottom topography, *Journal of Computational Physics* **222**, 2, pp. 720–739, doi:https://doi.org/10.1016/j.jcp.2006.08.014.

Gao, J., Ji, C., Liu, Y., Ma, X., and Gaidai, O. (2018). Numerical study on transient harbor oscillations induced by successive solitary waves, *Ocean Dynamics* **68**, 2, pp. 193–209, doi:https://doi.org/10.1007/s10236-017-1121-9.

Glaister, P. (1988). Approximate Riemann solutions of the shallow water equations, *Journal of Hydraulic Research* **26**, 3, pp. 293–306, doi:https://doi.org/10.1080/00221688809499213.

Gobbi, M. F., Kirby, J. T., and Wei, G. (2000). A fully nonlinear Boussinesq model for surface waves. Part 2. Extension to $O(kh)^4$, *Journal of Fluid Mechanics* **405**, pp. 181–210, doi:https://doi.org/10.1017/S0022112099007247.

Gómez-Gesteira, M., Crespo, A. J. C., Rogers, B. D., Dalrymple, R. A., Dominguez, J. M., and Barreiro, A. (2012a). SPHysics–Development of a free-surface fluid solver – Part 2: Efficiency and test cases, *Computers & Geosciences* **48**, pp. 300–307, doi:https://doi.org/10.1016/j.cageo.2012.02.028.

Gomez-Gesteira, M., Rogers, B. D., Crespo, A. J. C., Dalrymple, R. A., Narayanaswamy, M., and Dominguez, J. M. (2012b). SPHysics–Development of a free-surface fluid solver – Part 1: Theory and formulations. *Computers & Geosciences* **48**, pp. 289–299, doi:https://doi.org/10.1016/j.cageo.2012.02.029.

Gotoh, H. and Khayyer, A. (2018). On the state-of-the-art of particle methods for coastal and ocean engineering, *Coastal Engineering Journal* **60**, 1, pp. 79–103, doi:https://doi.org/10.1080/21664250.2018.1436243.

Gouin, M., Ducrozet, G., and Ferrant, P. (2014). Development of a highly nonlinear model for wave propagation over a variable bathymetry, in *Proceedings of the 26th International Workshop on Water Waves and Floating Bodies*.

Gouin, M., Ducrozet, G., and Ferrant, P. (2015). Validation of a nonlinear spectral model for water waves over a variable bathymetry, in *Proceedings of the 30th International Workshop on Water Waves and Floating Bodies*.

Gracia Garcia, V., García León, M., Sánchez-Arcilla Conejo, A., Gault, J., Oller, P., Fernández, J., Sairouni, A., Cristofori, E., and Toldrà, R. (2014). A new generation of early warning systems for coastal risk: The iCoast project, in *Proceedings of 34th International Conference on Coastal Engineering*, doi:https://doi.org/10.9753/icce.v34.management.18.

Green, A. E., Laws, N., and Naghdi, P. M. (1974). On the theory of water waves, *Proceedings of the Royal Society A: Maths, Physics, Engineering and Science* **338**, pp. 43–55, doi:https://doi.org/10.1098/rspa.1974.0072.

Green, A. E. and Naghdi, P. M. (1976a). A derivation of equations for wave propagation in water of variable depth, *Journal of Fluid Mechanics* **78**, pp. 237–246, doi:https://doi.org/10.1017/S0022112076002425.

Green, A. E. and Naghdi, P. M. (1976b). Directed fluid sheets, *Proceedings of the Royal Society A: Maths, Physics, Engineering and Science* **347**, pp. 447–473, doi:https://doi.org/10.1098/rspa.1976.0011.

Griffies, S. M., Adcroft, A. J., Banks, H., Böning, C. W., Chassignet, E. P., Danabasoglu, G., Danilov, S., Deleersnijder, E., Drange, H., England, M., *et al.* (2009). Problems and prospects in large-scale ocean circulation models, in *Proceedings of OceanObs'09*, pp. 410–431.

Grilli, A. R., Westcott, G., Grilli, S. T., Spaulding, M. L., Shi, F., and Kirby, J. T. (2020). Assessing coastal hazard from extreme storms with a phase resolving wave model: Case study of Narragansett, RI, USA, *Coastal Engineering* **160**, p. 103735, doi:https://doi.org/10.1016/j.coastaleng.2020.103735.

Grilli, S. T., Guyenne, P., and Dias, F. (2001). A fully non-linear model for three-dimensional overturning waves over an arbitrary bottom, *International Journal for Numerical Methods in Fluids* **35**, 7, pp. 829–867, doi:https://doi.org/10.1002/1097-0363(20010415)35:7⟨829::AID-FLD115⟩3.0.CO;2-2.

Grilli, S. T. and Horrillo, J. (1997). Numerical generation and absorption of fully nonlinear periodic waves, *Journal of Engineering Mechanics* **123**, 10, p. 1060–1069, doi:https://doi.org/10.1061/(ASCE)0733-9399(1997)123:10(1060).

Grilli, S. T., Skourup, J., and Svendsen, I. A. (1989). An efficient boundary element method for nonlinear water waves, *Engineering Analysis with Boundary Elements* **6**, 2, pp. 97–107, doi:https://doi.org/10.1016/0955-7997(89)90005-2.

Grilli, S. T. and Subramanya, R. (1996). Numerical modeling of wave breaking induced by fixed or moving boundaries, *Computational Mechanics* **17**, 6, pp. 374–391, doi:https://doi.org/10.1007/BF00363981.

Grilli, S. T., Vogelmann, S., and Watts, P. (2002). Development of a 3D numerical wave tank for modeling tsunami generation by underwater landslides. *Engineering Analysis with Boundary Elements* **26**, 4, pp. 301–313, doi: https://doi.org/10.1016/S0955-7997(01)00113-8.

Grue, J. (2005). A nonlinear model for surface waves interacting with a surface-piercing cylinder, in *Proceedings 20th International Workshop on Water Waves and Floating Bodies*, Longyearbyen, Norway, Vol. 29.

Guanche, R., Losada, I. J., and Lara, J. L. (2009). Numerical analysis of wave loads for coastal structure stability, *Coastal Engineering* **56**, 5–6, pp. 543–558, doi:https://doi.org/10.1016/j.coastaleng.2008.11.003.

Guéguen, Y., Dormieux, L., and Boutéca, M. (2004). Fundamentals of poromechanics, Chap. 1, *Mechanics of Fluid-Saturated Rocks*, Elsevier, Amsterdam, NL, **89**, pp. 1–54, doi:https://doi.org/10.1016/S0074-6142(03)80017-7.

Guignard, S., Grilli, S. T., Marcer, R., and Rey, V. (1999). Computation of shoaling and breaking waves in nearshore areas by the coupling of BEM and VOF methods, in *Proceedings of the 9th International Offshore and Polar Engineering Conference.*

Guimaraes, P. V., Farina, L., Toldo Jr, E., Diaz-Hernandez, G., and Akhmatskaya, E. (2015). Numerical simulation of extreme wave runup during storm events in Tramandaí Beach, Rio Grande do Sul, Brazil, *Coastal Engineering* **95**, pp. 171–180, doi:https://doi.org/10.1016/j.coastaleng.2014.10.008.

Gunstensen, A. K., Rothman, D. H., Zaleski, S., and Zanetti, G. (1991). Lattice Boltzmann model of immiscible fluids, *Physical Review A* **43**, 8, p. 4320, doi:https://doi.org/10.1103/PhysRevA.43.4320.

Guo, Z. and Shu, C. (2013). *Lattice Boltzmann Method and its Application in Engineering*, Vol. 3 (World Scientific, Singapore), doi:https://doi.org/10.1142/8806.

Guyenne, P. (2017). A high-order spectral method for nonlinear water waves in the presence of a linear shear current, *Computers & Fluids* **154**, pp. 224–235, doi:https://doi.org/10.1016/j.compfluid.2017.06.004.

Guyenne, P., Grilli, S. T., and Dias, F. (2000). Numerical modeling of fully nonlinear 3D overturning waves over arbitrary bottom, in *Proceedings of the Coastal Engineering Conference 2000*, pp. 417–428, doi:https://doi.org/10.1061/40549(276)33.

Guyenne, P. and Nicholls, D. P. (2008). A high-order spectral method for nonlinear water waves over moving bottom topography, *SIAM Journal on Scientific Computing* **30**, 1, pp. 81–101, doi:https://doi.org/10.1137/060666214.

Guyenne, P. and Părău, E. I. (2017). Numerical study of solitary wave attenuation in a fragmented ice sheet, *Physical Review Fluids* **2**, 3, p. 034002, doi: https://doi.org/10.1103/PhysRevFluids.2.034002.

Ha, T., Choi, J.-Y., Yoo, J., Chun, I., and Shim, J. (2014). Transformation of small-scale meteorological tsunami due to terrain complexity on the western coast of Korea, *Journal of Coastal Research* **70**, pp. 284–289, doi:https://doi.org/10.2112/SI70-048.1.

Hamidi, M. E., Hashemi, M. R., Talebbeydokhti, N., and Neill, S. P. (2012). Numerical modelling of the mild slope equation using localised differential quadrature method, *Ocean Engineering* **47**, pp. 88–103, doi:https://doi.org/10.1016/j.oceaneng.2012.03.004.

Hardy, J., Pomeau, Y., and De Pazzis, O. (1973). Time evolution of a two-dimensional classical lattice system, *Physical Review Letters* **31**, 5, doi:https://doi.org/10.1103/PhysRevLett.31.276.

Harlow, F. H. and Welch, J. E. (1965). Numerical calculation of time-dependent viscous incompressible flow of fluid with free surface, *The Physics of Fluids* **8**, 12, pp. 2182–2189, doi:https://doi.org/10.1063/1.1761178.

Harten, A., Lax, P. D., and Leer, B. V. (1983). On upstream differencing and Godunov-type schemes for hyperbolic conservation laws, *SIAM Review* **25**, 1, pp. 35–61, doi:https://doi.org/10.1137/1025002.

Hasselmann, K. (1962). On the non-linear energy transfer in a gravity-wave spectrum. Part 1. General theory, *Journal of Fluid Mechanics* **12**, 4, pp. 481–500, doi:https://doi.org/10.1017/S0022112062000373.

Hasselmann, K. (1974). On the spectral dissipation of ocean waves due to white capping, *Boundary-Layer Meteorology* **6**, 1, pp. 107–127, doi:https://doi.org/10.1007/BF00232479.

Hasselmann, K. F., Barnett, T. P., Bouws, E., Carlson, H., Cartwright, D. E., Enke, K., Ewing, J. A., Gienapp, H., Hasselmann, D. E., Kruseman, P., Meerburg, A., Müller, P., Olbers, D. J., Richter, K., Sell, W., and Walden, H. (1973). Measurements of wind-wave growth and swell decay during the Joint North Sea Wave Project (JONSWAP), *Ergaenzungsheft zur Deutschen Hydrographischen Zeitschrift, Reihe A – Supplement to the German Hydrographic Journal, Series A*, Vol. 12, doi:http://resolver.tudelft.nl/uuid:f204e188-13b9-49d8-a6dc-4fb7c20562fc.

Hasselmann, S. and Hasselmann, K. (1985). Computations and parameterizations of the nonlinear energy transfer in a gravity-wave spectrum. Part I: A new method for efficient computations of the exact nonlinear transfer integral, *Journal of Physical Oceanography* **15**, 11, pp. 1369–1377, doi:https://doi.org/10.1175/1520-0485(1985)015⟨1369:CAPOTN⟩2.0.CO;2.

Hasselmann, S., Hasselmann, K. F., Allender, J. H., and Barnett, T. P. (1985). Computations and parameterizations of the nonlinear energy transfer in a gravity-wave spectrum. Part II: Parameterizations of the nonlinear energy transfer for application in wave models, *Journal of Physical Oceanography* **15**, 11, pp. 1378–1391, doi:https://doi.org/10.1175/1520-0485(1985)015⟨1378:CAPOTN⟩2.0.CO;2.

He, X., Chen, S., and Zhang, R. (1999). A lattice Boltzmann scheme for incompressible multiphase flow and its application in simulation of Rayleigh–Taylor instability, *Journal of Computational Physics* **152**, 2, pp. 642–663, doi:https://doi.org/10.1006/jcph.1999.6257.

Heinze, C. (2003). Nonlinear hydrodynamic effects on fixed and oscillating structures in waves, PhD thesis, Oxford University, UK.

Hemavathi, S. and Manjula, R. (2019). Numerical modeling for wave attenuation by coastal vegetation using FLOW3D, *International Journal of Recent Technology and Engineering* **7**, 6S.

Henderson, K. L., Peregrine, D. H., and Dold, J. W. (1999). Unsteady water wave modulations: Fully nonlinear solutions and comparison with the nonlinear Schrödinger equation, *Wave Motion* **29**, 4, pp. 341–361, doi:https://doi.org/10.1016/S0165-2125(98)00045-6.

Herbers, T. H. C. and Burton, M. C. (1997). Nonlinear shoaling of directionally spread waves on a beach, *Journal of Geophysical Research: Oceans* **102**, C9, pp. 21101–21114, doi:https://doi.org/10.1029/97JC01581.

Hermans, A. (2000). A boundary element method for the interaction of free-surface waves with a very large floating flexible platform, *Journal of Fluids and Structures* **14**, 7, pp. 943–956, doi:https://doi.org/10.1006/jfls.2000.0313.

Hersbach, H. and Janssen, P. A. E. M. (1999). Improvement of the short-fetch behavior in the Wave Ocean Model (WAM), *Journal of Atmospheric and Oceanic Technology* **16**, 7, pp. 884–892, doi:https://doi.org/10.1175/1520-0426(1999)016⟨0884:IOTSFB⟩2.0.CO;2.

Hewitt, S., Margetts, L., Revell, A., Pankaj, P., and Levrero-Florencio, F. (2019). OpenFPCI: A parallel fluid–structure interaction framework, *Computer Physics Communications* **244**, pp. 469–482, doi:https://doi.org/10.1016/j.cpc.2019.05.016.

Higuera, P. (2015). Application of computational fluid dynamics to wave action on structures, PhD thesis, University of Cantabria, Spain.

Higuera, P. (2017). Olaflow: CFD for waves [Software]. doi:https://doi.org/10.5281/zenodo.1297013.

Higuera, P. (2020). Enhancing active wave absorption in RANS models, *Applied Ocean Research* **94**, doi:https://doi.org/10.1016/j.apor.2019.102000.

Higuera, P. (2021). Wave and structure interaction. Porous coastal structures., Chap. 6, in D. M. Kelly, A. Dimakopoulos, P. Higuera, *Advanced Numerical Modelling of Wave Structure Interactions* (CRC Press, Florida, USA), doi: https://doi.org/10.1201/9781351119542.

Higuera, P., Buldakov, E., and Stagonas, D. (2018a). Numerical modelling of wave interaction with an FPSO Using a Combination of OpenFOAM and Lagrangian models, in *Proceedings of the 28th International Ocean and Polar Engineering Conference*.

Higuera, P., Buldakov, E., and Stagonas, D. (2021). Simulation of steep waves interacting with a cylinder by coupling CFD and Lagrangian models, *International Journal of Offshore and Polar Engineering* **31**, 01, pp. 87–94, doi: https://doi.org/10.17736/ijope.2021.jc821.

Higuera, P., del Jesus, M., Lara, J. L., Losada, I. J., Guanche, Y., and Barajas, G. (2013a). Numerical simulation of three-dimensional breaking waves on a gravel slope using a two-phase flow Navier-Stokes model, *Journal of Computational and Applied Mathematics* **246**, pp. 144–152, doi:https://doi.org/10.1016/j.cam.2012.10.007.

Higuera, P., Lara, J. L., and Losada, I. J. (2013b). Realistic wave generation and active wave absorption for Navier-Stokes models: Application to Open-FOAM, *Coastal Engineering* **71**, pp. 102–118, doi:https://doi.org/10.1016/j.coastaleng.2012.07.002.

Higuera, P., Lara, J. L., and Losada, I. J. (2013c). Simulating coastal engineering processes with OpenFOAM, *Coastal Engineering* **71**, pp. 119–134, doi: https://doi.org/10.1016/j.coastaleng.2012.06.002.

Higuera, P., Lara, J. L., and Losada, I. J. (2014a). Three-dimensional interaction of waves and porous coastal structures using OpenFOAM. Part I: Formulation and validation, *Coastal Engineering* **83**, pp. 243–258, doi: https://doi.org/10.1016/j.coastaleng.2013.08.010.

Higuera, P., Lara, J. L., and Losada, I. J. (2014b). Three-dimensional interaction of waves and porous coastal structures using OpenFOAM. Part II: Applications, *Coastal Engineering* **83**, pp. 259–270, doi:https://doi.org/10.1016/j.coastaleng.2013.09.002.

Higuera, P., Liu, P. L. F., Lin, C., Wong, W. Y., and Kao, M. J. (2018b). Laboratory-scale swash flows generated by a non-breaking solitary wave on a steep slope, *Journal of Fluid Mechanics* **847**, pp. 186–227, doi: https://doi.org/10.1017/jfm.2018.321.

Hirai, T., Sou, A., and Nihei, Y. (2018). Wave load acting on advanced spar in regular waves, in *International Conference on Offshore Mechanics and Arctic Engineering*, Vol. 51258, doi:https://doi.org/10.1115/OMAE2018-77821.

Hirt, C. W., Amsden, A. A., and Cook, J. L. (1974). An arbitrary Lagrangian–Eulerian computing method for all flow speeds, *Journal of Computational Physics* **14**, 3, pp. 227–253, doi:https://doi.org/10.1016/0021-9991(74)90051-5.

Hirt, C. W. and Nichols, B. D. (1981). Volume of fluid (VOF) method for the dynamics of free boundaries, *Journal of Computational Physics* **39**, pp. 201–225, doi:https://doi.org/10.1016/0021-9991(81)90145-5.

Ho, M. (2019). *Modeling and Validation of Coastal Wastewater Effluent Plumes using High-Resolution Nonhydrostatic Regional Ocean Modeling System.* (University of California, Los Angeles).

Holthuijsen, L. H. (2010). *Waves in Oceanic and Coastal Waters.* (Cambridge University Press, UK).

Holthuijsen, L. H., Herman, A., and Booij, N. (2003). Phase-decoupled refraction-diffraction for spectral wave models, *Coastal Engineering* **49**, 4, pp. 291–305, doi:https://doi.org/10.1016/S0378-3839(03)00065-6.

Hong, G. W. (1996). Mathematical models for combined refraction-diffraction of waves on non-uniform current and depth, *China Ocean Engineering* **10**, 4, pp. 433–454.

Hsiao, S. C., Liu, P. L., and Chen, Y. (2002). Nonlinear water waves propagating over a permeable bed, *Proceedings of the Royal Society A: Maths, Physics, Engineering and Science* **458**, 2022, pp. 1291–1322, doi:https://doi.org/10.1098/rspa.2001.0903.

Hsiao, S. S., Chang, C. M., and Wen, C. C. (2010). An analytical solution to the modified mild-slope equation for waves propagating around a circular conical island, *Journal of Marine Science and Technology* **18**, 4, pp. 520–529, doi:https://doi.org/10.51400/2709-6998.1905.

Hsu, T.-W., Ou, S.-H., and Liau, J.-M. (2005). Hindcasting nearshore wind waves using a FEM code for SWAN, *Coastal Engineering* **52**, 2, pp. 177–195, doi: https://doi.org/10.1016/j.coastaleng.2004.11.005.

Hu, K., Mingham, C. G., and Causon, D. M. (2000). Numerical simulation of wave overtopping of coastal structures using the non-linear shallow water equations, *Coastal Engineering* **41**, 4, pp. 433–465, doi:https://doi.org/10.1016/S0378-3839(00)00040-5.

Hu, L., Dong, Z., Peng, C., and Wang, L.-P. (2021). Direct numerical simulation of sediment transport in turbulent open channel flow using the Lattice Boltzmann method, *Fluids* **6**, 6, p. 217, doi:https://doi.org/10.3390/fluids6060217.

Hu, P., Wu, G. X., and Ma, Q. W. (2002). Numerical simulation of nonlinear wave radiation by a moving vertical cylinder, *Ocean Engineering* **29**, 14, pp. 1733–1750, doi:https://doi.org/10.1016/S0029-8018(02)00002-1.

Huang, H., Ding, P. X., and Lu, X. H. (2000). Mild-slope equation for water waves propagating over non-uniform currents and uneven bottoms, *Acta Oceanologica Sinica* **19**, 3, pp. 23–31.

Hubbard, M. E. and Dodd, N. (2002). A 2D numerical model of wave run-up and overtopping, *Coastal Engineering* **47**, 1, pp. 1–26, doi:https://doi.org/10.1016/S0378-3839(02)00094-7.

Hughes, T. J. R., Liu, W. K., and Zimmermann, T. K. (1981). Lagrangian–Eulerian finite element formulation for incompressible viscous flows, *Computer Methods in Applied Mechanics and Engineering* **29**, 3, pp. 329–349, doi:https://doi.org/10.1016/0045-7825(81)90049-9.

Ioualalen, M., Asavanant, J., Kaewbanjak, N., Grilli, S., Kirby, J., and Watts, P. (2007). Modeling the 26 December 2004 Indian Ocean tsunami: Case study of impact in Thailand, *Journal of Geophysical Research: Oceans* **112**, C7, doi:https://doi.org/10.1029/2006JC003850.

Isobe, M. (1986). A parabolic refraction-diffraction equation in the ray-front coordinate system, *Coastal Engineering 1986 (Proceeding of the International Conference on Coastal Engineering)*, pp. 306–317, doi:https://doi.org/10.1061/9780872626003.024.

Isobe, M. (1987). A parabolic equation model for transformation of irregular waves due to refraction, diffraction and breaking, *Coastal Engineering in Japan* **30**, 1, pp. 33–47, doi:https://doi.org/10.1080/05785634.1987.11924463.

Jacobsen, N. G. (2017). *waves2foam Manual* (Deltares, The Netherlands).

Jacobsen, N. G., Fuhrman, D. R., and Fredsøe, J. (2012). A wave generation toolbox for the open-source CFD library: OpenFOAM, *International Journal for Numerical Methods in Fluids.* **70**, 9, pp. 1073–1088, doi:https://doi.org/10.1002/fld.2726.

Jacobsen, N. G. and McFall, B. C. (2022). Wave-averaged properties for non-breaking waves in a canopy: Viscous boundary layer and vertical shear stress distribution, *Coastal Engineering* **174**, doi:https://doi.org/10.1016/j.coastaleng.2022.104117.

Janssen, P. (1986). *On the effects of gustiness on wave growth*, KNMI Afdeling Oceanografisch Onderzoek Memo – KNMI, NL, Department of Oceanographic Research Memo.

Janssen, P. and Janssen, P. A. (2004). *The Interaction of Ocean Waves and Wind* (Cambridge University Press, UK), doi:https://doi.org/10.1017/CBO9780511525018.

Janssen, P. A. (1982). Quasilinear approximation for the spectrum of wind-generated water waves, *Journal of Fluid Mechanics* **117**, pp. 493–506, doi: https://doi.org/10.1017/S0022112082001736.

Janssen, P. A. (1989). Wave-induced stress and the drag of air flow over sea waves, *Journal of Physical Oceanography* **19**, 6, pp. 745–754, doi:https://doi.org/10.1175/1520-0485(1989)019¡0745:WISATD¿2.0.CO;2.

Janssen, P. A. (1992). Experimental evidence of the effect of surface waves on the airflow, *Journal of Physical Oceanography* **22**, 12, pp. 1600–1604, doi: https://doi.org/10.1175/1520-0485(1992)022⟨1600:EEOTEO⟩2.0.CO;2.

Janssen, P. A. and Onorato, M. (2007). The intermediate water depth limit of the Zakharov equation and consequences for wave prediction, *Journal of Physical Oceanography* **37**, 10, pp. 2389–2400, doi:https://doi.org/10.1175/JPO3128.1.

Jeffreys, H. (1925). On the formation of water waves by wind, *Proceedings of the Royal Society of London. Series A* **107**, 742, pp. 189–206, doi:https://doi.org/10.1098/rspa.1925.0015.

Jeffreys, H. (1926). On the formation of water waves by wind (second paper), *Proceedings of the Royal Society of London. Series A* **110**, 754, pp. 241–247, doi:https://doi.org/10.1098/rspa.1926.0014.

Jelesnianski, C. P. (1992). *SLOSH: Sea, Lake, and Overland Surges from Hurricanes*, Vol. 48 (US Department of Commerce, National Oceanic and Atmospheric Administration Technical Report NWS 48).

Jeng, D., Cha, D., Lin, Y., and Hu, P. (2001). Wave-induced pore pressure around a composite breakwater, *Ocean Engineering* **28**, 10, pp. 1413–1435, doi: https://doi.org/10.1016/S0029-8018(00)00059-7.

Jeng, D.-S. and Ou, J. (2010). 3D models for wave-induced pore pressures near breakwater heads, *Acta Mechanica* **215**, 1, pp. 85–104, doi:https://doi.org/10.1007/s00707-010-0303-z.

Jeng, D.-S. and Ye, J. (2012). Three-dimensional consolidation of a porous unsaturated seabed under rubble mound breakwater, *Ocean Engineering* **53**, pp. 48–59, doi:https://doi.org/10.1016/j.oceaneng.2012.06.004.

Jensen, B., Jacobsen, N. G., and Christensen, E. D. (2014). Investigations on the porous media equations and resistance coefficients for coastal structures, *Coastal Engineering* **84**, pp. 56–72, doi:https://doi.org/10.1016/j.coastaleng.2013.11.004.

Jin, J. and Meng, B. (2011). Computation of wave loads on the superstructures of coastal highway bridges, *Ocean Engineering* **38**, 17–18, pp. 2185–2200, doi:https://doi.org/10.1016/j.oceaneng.2011.09.029.

Kaihatu, J. M. (1997). Review and verification of numerical wave models for near coastal areas – Part 1: Review of mild slope equation. Relevant approximations and technical details of numerical wave models, Defense Technical Information Center, USA.

Kaihatu, J. M. (2001). Improvement of parabolic nonlinear dispersive wave model, *Journal of Waterway, Port, Coastal and Ocean Engineering* **127**, 2, pp. 113–121, doi:https://doi.org/10.1061/(ASCE)0733-950X(2001)127: 2(113).

Kantorovich, L. V. and Krylov, V. I. (1958). *Approximate Methods of Higher Analysis* (P. Noordhoff Ltd., Groningen, NL), Groningen, NL, doi:https://doi.org/10.2307/3612589.

Karniadakis, G. and Sherwin, S. (2013). *Spectral/HP Element Methods for Computational Fluid Dynamics* (Oxford University Press, UK), doi:https://doi.org/10.1093/acprof:oso/9780198528692.001.0001.

Katsidoniotaki, E. and Göteman, M. (2020). Comparison of dynamic mesh methods in OpenFOAM for a WEC in extreme waves, in Carlos Guedes Soares *Developments in Renewable Energies Offshore* (CRC Press, Florida, USA), pp. 214–222, doi:https://doi.org/10.1201/9781003134572.

Keller, J. B. (1958). Surface waves on water of non-uniform depth, *Journal of Fluid Mechanics* **4**, pp. 607–614, doi:https://doi.org/10.1017/S0022112058000690.

Kennedy, A. B., Chen, Q., Kirby, J. T., and Dalrymple, R. A. (2000). Boussinesq modeling of wave transformation, breaking, and runup. I: 1D, *Journal of Waterway, Port, Coastal, and Ocean Engineering* **126**, doi:https://doi.org/10.1061/(ASCE)0733-950X(2000)126:1(39).

Kennedy, A. B., Kirby, J. T., Chen, Q., and Dalrymple, R. A. (2001). Boussinesq-type equations with improved nonlinear performance, *Wave Motion* **33**, 3, pp. 225–243, doi:https://doi.org/10.1016/S0165-2125(00)00071-8.

Khanh, L. P. (2019). Wave attenuation in coastal mangroves: Mangrove squeeze in the Mekong delta, PhD thesis, Delft University of Technology, Delft.

Kharif, C., Abid, M., and Touboul, J. (2017). Rogue waves in shallow water in the presence of a vertically sheared current, *Journal of Ocean Engineering and Marine Energy* **3**, pp. 301–308, doi:https://doi.org/10.1007/s40722-017-0085-7.

Kharif, C., Giovanangeli, J.-P., Touboul, J., Grare, L., and Pelinovsky, E. (2008). Influence of wind on extreme wave events: Experimental and numerical approaches, *Journal of Fluid Mechanics* **594**, pp. 209–247, doi:https://doi.org/10.1017/S0022112007009019.

Khayyer, A., Gotoh, H., Shimizu, Y., Nishijima, Y., and Nakano, A. (2020a). 3D MPS-MPS coupled FSI solver for simulation of hydroelastic fluid-structure interactions in coastal engineering, *Journal of Japan Society of Civil Engineers, Ser. B2 (Coastal Engineering)* **76**, 2, doi:https://doi.org/10.2208/kaigan.76.2_I_37.

Khayyer, A., Gotoh, H., Shimizu, Y., Sasagawa, H., Nakano, A., (2020b). A 3D fully Lagrangian meshfree hydroelastic solver; 3D ISPH-SPH, in *Proceedings of the 30th International Ocean and Polar Engineering Conference.*

Kim, D. and Kim, M. (1997). Wave-current-body interaction by a tune-domain high-order boundary element method, in *Proceedings of the 7th International Ocean and Polar Engineering Conference*, Honolulu, Hawaii, USA.

Kim, D.-H. (2015). H2D morphodynamic model considering wave, current and sediment interaction. *Coastal Engineering* **95**, pp. 20–34, doi:https://doi.org/10.1016/j.coastaleng.2014.09.006.

Kim, D.-H. and Lynett, P. J. (2011). Turbulent mixing and passive scalar transport in shallow flows, *Physics of Fluids* **23**, 1, p. 016603, doi:https://doi.org/10.1063/1.3531716.

Kim, D. H., Lynett, P. J., and Socolofsky, S. A. (2009). A depth-integrated model for weakly dispersive, turbulent, and rotational fluid flows, *Ocean Modelling* **27**, pp. 198–214, doi:https://doi.org/10.1016/j.ocemod.2009.01.005.

Kim, S.-H., Yamashiro, M., and Yoshida, A. (2010). A simple two-way coupling method of BEM and VOF model for random wave calculations, *Coastal Engineering* **57**, 11–12, pp. 1018–1028, doi:https://doi.org/10.1016/j.coastaleng.2010.06.006.

Kirby, J. T. (1984). A note on linear surface wave-current interaction, *Journal of Geophysical Research* **89**, pp. 745–747, doi:https://doi.org/10.1029/JC089iC01p00745.

Kirby, J. T. (1986). A general wave equation for waves over rippled beds, *Journal of Fluid Mechanics* **162**, pp. 171–186, doi:https://doi.org/10.1017/S0022112086001994.

Kirby, J. T. (1986a). Higher-order approximations in the parabolic equation method for water waves, *Journal of Geophysical Research: Oceans* **91**, C1, pp. 933–952, doi:https://doi.org/10.1029/JC091iC01p00933.

Kirby, J. T. (1986b). Rational approximations in the parabolic equation method for water waves, *Coastal Engineering* **10**, 4, pp. 355–378, doi:https://doi.org/10.1016/0378-3839(86)90021-9.

Kirby, J. T. (1988). Parabolic wave computations in non-orthogonal coordinate systems, *Journal of Waterway, Port, Coastal and Ocean Engineering* **114**, 6, pp. 673–685, doi:https://doi.org/10.1061/(ASCE)0733-950X(1988)114:6(673).

Kirby, J. T. (2016). Boussinesq models and their application to coastal processes across a wide range of scales, *Journal of Waterway, Port, Coastal, and Ocean Engineering* **142**, 6, doi:https://doi.org/10.1061/(ASCE)WW.1943-5460.0000350.

Kirby, J. T. and Dalrymple, R. A. (1983). A parabolic equation for the combined refraction–diffraction of stokes waves by mildly varying topography, *Journal of Fluid Mechanics* **136**, pp. 453–466, doi:https://doi.org/10.1017/S0022112083002232.

Kirby, J. T. and Dalrymple, R. A. (1984). Verification of a parabolic equation for propagation of weakly-nonlinear waves, *Coastal Engineering* **8**, 3, pp. 219–232, doi:https://doi.org/10.1016/0378-3839(84)90002-4.

Kirby, J. T. and Dalrymple, R. A. (1986). An approximate model for nonlinear dispersion in monochromatic wave propagation models, *Coastal Engineering* **9**, 6, pp. 545–561, doi:https://doi.org/10.1016/0378-3839(86)90003-7.

Kirby, J. T. and Dalrymple, R. A. (1994). Combined refraction/diffraction model, University of Delaware, Newark, CACR Report No.92–04.

Kirby, J. T., Dalrymple, R. A., and Kaku, H. (1994). Parabolic approximations for water waves in conformal coordinate systems, *Coastal Engineering* **23**, 3–4, pp. 185–213, doi:https://doi.org/10.1016/0378-3839(94)90001-9.

Kirby, J. T., Wei, G., Chen, Q., Kennedy, A. B., and Dalrymple, R. A. (1998). FUNWAVE 1.0: Fully nonlinear Boussinesq wave model – Documentation and user's manual, research report NO. CACR-98-06, Center for Applied Coastal Research, Delaware, USA.

Kirstetter, G., Delestre, O., Lagrée, P.-Y., Popinet, S., and Josserand, C. (2021). B-flood 1.0: An open-source saint-venant model for flash-flood simulation using adaptive refinement, *Geoscientific Model Development* **14**, 11, pp. 7117–7132.

Ko, D. H., Jeong, S. T., and Oh, N. S. (2015). Numerical simulation test of scour around offshore jacket structure using FLOW-3D, *Journal of Korean Society of Coastal and Ocean Engineers* **27**, 6, pp. 373–381, doi:https://doi.org/10.9765/KSCOE.2015.27.6.373.

Koçyigit, M. B., Falconer, R. A., and Lin, B. (2002). Three-dimensional numerical modelling of free surface flows with non-hydrostatic pressure, *International Journal for Numerical Methods in Fluids* **40**, 9, pp. 1145–1162, doi:https://doi.org/10.1002/fld.376.

Komen, G. J., Cavaleri, L., Donelan, M., Hasselmann, K., Hasselmann, S., and Janssen, P. A. E. M. (1996). *Dynamics and Modelling of Ocean Waves* (Cambridge University Press, UK), doi:https://doi.org/10.1017/CBO9780511628955.

Komen, G. J., Hasselmann, S., and Hasselmann, K. (1984). On the existence of a fully developed wind-sea spectrum, *Journal of Physical Oceanography* **14**, 8, pp. 1271–1285, doi:https://doi.org/10.1175/1520-0485(1984)014⟨1271: OTEOAF⟩2.0.CO;2.

Koshizuka, S. (1995). A particle method for incompressible viscous flow with fluid fragmentation, *Computational Fluid Dynamics Journal* **4**, p. 29.

Kostense, J. K., Dingemans, M. W., and Van den Bosch, P. (1989). Wave-current interaction in harbours, *Coastal Engineering 1988 (Proceeding of the International Conference on Coastal Engineering)* pp. 32–46 doi:https://doi.org/10.1061/9780872626874.003.

Krüger, T., Kusumaatmaja, H., Kuzmin, A., Shardt, O., Silva, G., and Viggen, E. M. (2017). *The Lattice Boltzmann Method* (Springer International Publishing, Berlin, Germany), doi:https://doi.org/10.1007/978-3-319-44649-3.

Kubo, Y., Kotake, Y., Isobe, M., and Watanabe, A. (1993). Time-dependent mild slope equation for random waves, *Coastal Engineering 1992 (Proceeding of the International Conference on Coastal Engineering)* pp. 419–431 doi: https://doi.org/10.1061/9780872629332.031.

Kurganov, A. (2018). Finite-volume schemes for shallow-water equations, *Acta Numerica* **27**, pp. 289–351, doi:https://doi.org/10.1017/S09624929 18000028.

Lahoz, M. G. and Albiach, J. C. C. (1997). A two-way nesting procedure for the WAM model: Application to the Spanish coast, *Journal of Offshore*

Mechanics and Arctic Engineering **119**, pp. 20–24, doi:https://doi.org/10. 1115/1.2829040.

Lai, Z., Chen, C., Cowles, G. W., and Beardsley, R. C. (2010). A nonhydro-static version of FVCOM: 1. Validation experiments, *Journal of Geophysical Research: Oceans* **115**, C11, doi:https://doi.org/10.1029/2009JC005525.

Lannes, D. and Bonneton, P. (2009). Derivation of asymptotic two-dimensional time-dependent equations for surface water wave propagation, *Physics of Fluids* **21**, 1, p. 16601, doi:https://doi.org/10.1063/1.3053183.

Lannes, D. and Marche, F. (2015). A new class of fully nonlinear and weakly dispersive Green–Naghdi models for efficient 2D simulations, *Journal of Computational Physics* **282**, pp. 238–268, doi:http://doi.org/10.1016/j.jcp. 2014.11.016.

Lara, J. L., del Jesus, M., and Losada, I. J. (2012). Three-dimensional interaction of waves and porous structures. Part II: Model validation, *Coastal Engineering* **64**, pp. 26–46, doi:https://doi.org/10.1016/j.coastaleng.2012.01.009.

Lara, J. L., Garcia, N., and Losada, I. J. (2006). RANS modelling applied to random wave interaction with submerged permeable structures, *Coastal Engineering* **53**, pp. 395–417, doi:https://doi.org/10.1016/j.coastaleng.2005.11. 003.

Lara, J. L., Losada, I. J., and Guanche, R. (2008). Wave interaction with low mound breakwaters using a RANS model, *Ocean Engineering* **35**, pp. 1388–1400, doi:https://doi.org/10.1016/j.oceaneng.2008.05.006.

Larsen, B. E. and Fuhrman, D. R. (2018). On the over-production of turbulence beneath surface waves in Reynolds-averaged Navier–Stokes models, *Journal of Fluid Mechanics* **853**, pp. 419–460, doi:https://doi.org/10.1017/jfm. 2018.577.

Larsen, B. E., Fuhrman, D. R., and Sumer, B. M. (2016). Simulation of wave-plus-current scour beneath submarine pipelines, *Journal of Waterway, Port, Coastal, and Ocean Engineering* **142**, 5, doi:https://doi.org/10.1061/ (ASCE)WW.1943-5460.0000338.

Latham, J.-P., Munjiza, A., Mindel, J., Xiang, J., Guises, R., Garcia, X., Pain, C., Gorman, G., and Piggott, M. (2008). Modelling of massive particulates for breakwater engineering using coupled FEMDEM and CFD, *Particuology* **6**, 6, pp. 572–583, doi:https://doi.org/10.1016/j.partic.2008.07.010.

Lavrenov, I., Resio, D., and Zakharov, V. (2001). Numerical simulation of weak turbulent Kolmogorov spectrum in water surface waves, Tech. rep., Arctic and Antarctic Research Institute, Saint Petersburg, Russia.

Le Métayer, O., Gavrilyuk, S., and Hank, S. (2010). A numerical scheme for the Green–Naghdi model, *Journal of Computational Physics* **229**, 6, pp. 2034–2045, doi:https://doi.org/10.1016/j.jcp.2009.11.021.

Lee, C., Park, W. S., Cho, Y. S., and Suh, K. D. (1998). Hyperbolic mild-slope equations extended to account for rapidly varying topography, *Coastal Engineering* **34**, 3–4, pp. 243–257, doi:https://doi.org/10.1016/ S0378-3839(98)00028-3.

Lee, K.-H., Bae, J.-H., Jung, U. J., Choi, G.-H., and Kim, D.-S. (2019). Numerical simulation of nonlinear interaction between composite breakwater and seabed under irregular wave action by olaFlow model, *Journal of*

Korean Society of Coastal and Ocean Engineers **31**, 3, pp. 129–145, doi: https://doi.org/10.9765/KSCOE.2019.31.3.129.

Lesser, G. R., Roelvink, J. v., van Kester, J. T. M., and Stelling, G. (2004). Development and validation of a three-dimensional morphological model, *Coastal Engineering* **51**, 8–9, pp. 883–915, doi:https://doi.org/10.1016/j.coastaleng.2004.07.014.

LeVeque, R. J. (1998). Balancing source terms and flux gradients in high-resolution Godunov methods: The quasi-steady wave-propagation algorithm, *Journal of Computational Physics* **146**, 1, pp. 346–365, doi:https://doi.org/10.1006/jcph.1998.6058.

Li, B. and Anastasiou, K. (1992). Efficient elliptic solvers for the mild-slope equation using the multigrid technique, *Coastal Engineering* **16**, 3, pp. 245–266, doi:https://doi.org/10.1016/0378-3839(92)90044-U.

Li, B. and Fleming, C. A. (1997). A three dimensional multigrid model for fully nonlinear water waves, *Coastal Engineering* **30**, 3–4, pp. 235–258, doi:https://doi.org/10.1016/S0378-3839(96)00046-4.

Li, B., Reeve, D. E., and Fleming, C. A. (1993). Numerical solution of the elliptic mild-slope equation for irregular wave propagation, *Coastal Engineering* **20**, 1–2, pp. 85–100, doi:https://doi.org/10.1016/0378-3839(93)90056-E.

Li, C., Xu, C., Gui, C., and Fox, M. D. (2005). Level set evolution without re-initialization: A new variational formulation, in *Proceedings of the 2005 IEEE Computer Society Conference on Computer Vision and Pattern Recognition (CVPR'05)*, Vol. 1, pp. 430–436, doi:https://doi.org/10.1109/CVPR.2005.213.

Li, Q., Wang, J., Yan, S., Gong, J., and Ma, Q. (2018). A zonal hybrid approach coupling FNPT with OpenFOAM for modelling wave-structure interactions with action of current, *Ocean Systems Engineering* **8**, 4, pp. 381–407, doi: https://doi.org/10.12989/ose.2018.8.4.381.

Li, R., Jiang, S., and Jiang, B. (2010). Tide simulation using the mild-slope equation with Coriolis force and bottom friction, *Acta Oceanologica Sinica* **29**, 6, pp. 44–50, doi:https://doi.org/10.1007/s13131-010-0075-2.

Li, R. J. (2001). Wave transformation model considering the effect of nonlinear dispersion (in Chinese), *Acta Oceanologica Sinica* **23**, pp. 102–108.

Li, Y., Ong, M. C., and Fuhrman, D. R. (2020). CFD investigations of scour beneath a submarine pipeline with the effect of upward seepage, *Coastal Engineering* **156**, doi:https://doi.org/10.1016/j.coastaleng.2019.103624.

Li, Y. C. and Zhang, Y. G. (1995). Refraction-diffraction of irregular wave and directional spectrum with currents (in Chinese), *Journal of Hydrodynamics* **10**, 6.

Liang, Z. and Jeng, D.-S. (2021). PORO-FSSI-FOAM model for seafloor liquefaction around a pipeline under combined random wave and current loading, *Applied Ocean Research* **107**, doi:https://doi.org/10.1016/j.apor.2020.102497.

Lin, G. (2001). The parabolic mild-slope equation for irregular waves (in Chinese), *Journal of Hydraulic Engineering* **6**.

Lin, G. and Qiu, D. H. (2000). The variational principal and numerical simulation of parabolic mild-slope equation (in Chinese), *Acta Oceanologica Sinica* **22**, 1, pp. 125–130.

Lin, M.-C. and Hsiao, S.-S. (1994). Boundary element analysis of wave-current interaction around a large structure, *Engineering Analysis with Boundary Elements* **14**, 4, pp. 325–334, doi:https://doi.org/10.1016/0955-7997(94) 90062-0.

Lin, P. (2004). A compact numerical algorithm for solving the time-dependent mild slope equation, *International Journal for Numerical Methods in Fluids* **45**, 6, pp. 625–642, doi:https://doi.org/10.1002/fld.716.

Lin, P. (2008). *Numerical Modeling of Water Waves* (CRC Press, Florida, USA).

Lin, P. and Liu, P.-F. (1998). A numerical study of breaking waves in the surf zone, *Journal of Fluid Mechanics* **359**, pp. 239–264, doi:https://doi.org/10. 1017/S002211209700846X.

Lind, S. J., Rogers, B. D., and Stansby, P. K. (2020). Review of smoothed particle hydrodynamics: Towards converged Lagrangian flow modelling, *Proceedings of the Royal Society A* **476**, 2241, doi:https://doi.org/10.1098/rspa.2019. 0801.

Liu, G., Zhang, J., and Zhang, Q. (2021). A high-performance three-dimensional lattice Boltzmann solver for water waves with free surface capturing, *Coastal Engineering* **165**, p. 103865, doi:https://doi.org/10.1016/j.coastaleng.2021. 103865.

Liu, P. L.-F. (1983). Wave-current interactions on a slowly varying topography, *Journal of Geophysical Research: Oceans* **88**, C7, pp. 4421–4426, doi: https://doi.org/10.1029/JC088iC07p04421.

Liu, P. L.-F. (1986). Parabolic wave equation for surface water waves, Coastal Engineering Research Center Vicksburg, MS.

Liu, P. L.-F. (1995). Model equations for wave propagations from deep to shallow water, in P. L.-F. Liu (ed.), *Advances in Coastal and Ocean Engineering: (Volume 1)* (World Scientific, Singapore), pp. 125–157.

Liu, P. L.-F. and Boissevain, P. L. (1988). Wave propagation between two breakwaters, *Journal of Waterway, Port, Coastal and Ocean Engineering* **114**, 2, pp. 237–247, doi:https://doi.org/10.1061/(ASCE)0733-950X(1988)114: 2(237).

Liu, P. L.-F., Cho, Y. S., Briggs, M. J., Kanoglu, U., and Synolakis, C. E. (1995). Runup of solitary waves on a circular Island, *Journal of Fluid Mechanics* **302**, pp. 259–285, doi:https://doi.org/10.1017/S0022112095004095.

Liu, P. L.-F., Lin, P., Chang, K., and Sakakiyama, T. (1999). Numerical modeling of wave interaction with porous structures, *Journal of Waterway, Port, Coastal and Ocean Engineering* **125**, pp. 322–330, doi:https://doi.org/10. 1061/(ASCE)0733-950X(1999)125:6(322).

Liu, P. L.-F. and Losada, I. J. (2002). Wave propagation modeling in coastal engineering, *Journal of Hydraulic Research* **40**, 3, pp. 229–240, doi:https://doi.org/10.1080/00221680209499939.

Liu, P. L.-F. and Tsay, T. K. (1984). Refraction-diffraction model for weakly nonlinear water waves, *Journal of Fluid Mechanics* **141**, pp. 265–274, doi: https://doi.org/10.1017/S0022112084000835.

Liu, X. and García, M. H. (2008). Three-dimensional numerical model with free water surface and mesh deformation for local sediment scour, *Journal of Waterway, Port, Coastal, and Ocean Engineering* **134**, 4, pp. 203–217, doi: https://doi.org/10.1061/(ASCE)0733-950X(2008)134:4(203).

Liu, Y., Dommermuth, D. G., and Yue, D. K. (1992). A high-order spectral method for nonlinear wave–body interactions, *Journal of Fluid Mechanics* **245**, pp. 115–136, doi:https://doi.org/10.1017/S0022112092000375.

Liu, Y. and Yue, D. K. (1998). On generalized Bragg scattering of surface waves by bottom ripples, *Journal of Fluid Mechanics* **356**, pp. 297–326, doi:https://doi.org/10.1017/S0022112097007969.

Liu, Z. B., Fang, K. Z., and Cheng, Y. Z. (2018). A new multi-layer irrotational Boussinesq-type model for highly nonlinear and dispersive surface waves over a mildly sloping seabed, *Journal of Fluid Mechanics* **842**, pp. 323–353, doi:https://doi.org/10.1017/jfm.2018.99.

Longuet-Higgins, M. and Stewart, R. (1962). Radiation stresses and mass transport in gravity waves with applications to surf beats, *Journal of Fluid Mechanics* **13**, 4, pp. 529–562, doi:https://doi.org/10.1017/S0022112062000877.

Longuet-Higgins, M. S. (1969). On wave breaking and the equilibrium spectrum of wind-generated waves, *Proceedings of the Royal Society of London. A. Mathematical and Physical Sciences* **310**, 1501, pp. 151–159, doi:https://doi.org/10.1098/rspa.1969.0069.

Longuet-Higgins, M. S. and Cokelet, E. D. (1976). The deformation of steep surface waves on water - I. A numerical method of computation, *Proceedings of the Royal Society of London. A. Mathematical and Physical Sciences* **350**, 1660, pp. 1–26, doi:https://doi.org/10.1098/rspa.1976.0092.

López, J., Hernández, J., Gómez, P., and Faura, F. (2004). A volume of fluid method based on multidimensional advection and spline interface reconstruction, *Journal of Computational Physics* **195**, 2, pp. 718–742, doi:https://doi.org/10.1016/j.jcp.2003.10.030.

Losada, I. J., Lara, J. L., and del Jesus, M. (2016). Modeling the interaction of water waves with porous coastal structures, *Journal of Waterway, Port, Coastal and Ocean Engineering* **142**, 6, doi:https://doi.org/10.1061/(ASCE)WW.1943-5460.0000361.

Losada, I. J., Lara, J. L., Guanche, R., and Gonzalez-Ondina, J. M. (2008). Numerical analysis of wave overtopping of rubble mound breakwaters, *Coastal Engineering* **55**, 1, pp. 47–62, doi:https://doi.org/10.1016/j.coastaleng.2007.06.003.

Losada, I. J., Lara, J. L., Guanche, R., and Gonzalez-Ondina, J. M. (2009). Towards an engineering application of COBRAS (Cornell Breaking Wave

And Structures), in Patrick Lynett (ed.), *Nonlinear Wave Dynamics: Selected Papers of the Symposium Held in Honor of Philip L-F Liu's 60th Birthday* (World Scientific, Singapore), pp. 89–108, doi:https://doi.org/10.1142/9789812709042_0004.

Lou, J. and Massel, S. R. (1994). A combined refraction-diffraction-dissipation model of wave propagation, *Chinese Journal of Oceanology and Limnology* **12**, 4, pp. 361–371, doi:https://doi.org/10.1007/BF02850497.

Lozano, C. and Liu, P. L. F. (1980). Refraction–diffraction model for linear surface water waves, *Journal of Fluid Mechanics* **101**, 4, pp. 705–720, doi:https://doi.org/10.1017/S0022112080001887.

Luettich, R. A., Westerink, J. J., Scheffner, N. W. (1992). ADCIRC: An advanced three-dimensional circulation model for shelves, coasts, and estuaries, Report 1, Theory and methodology of ADCIRC-2DD1 and ADCIRC-3DL. Tech. rep., Coastal Engineering Research Center (US).

Luo, W. and Monbaliu, J. (1994). Effects of the bottom friction formulation on the energy balance for gravity waves in shallow water, *Journal of Geophysical Research: Oceans* **99**, C9, pp. 18501–18511, doi:https://doi.org/10.1029/94JC01230.

Lynett, P. and Liu, P. L.-F. (2002). A numerical study of submarine landslide generated waves and runup, *Proceedings of the Royal Society A* **458**, pp. 2885–2910, doi:https://doi.org/10.1098/rspa.2002.0973.

Lynett, P. and Liu, P. L.-F. (2004a). A two-layer approach to wave modelling, *Proceedings of the Royal Society A* **460**, pp. 2637–2669, doi:https://doi.org/10.1098/rspa.2004.1305.

Lynett, P. J. (2006). Nearshore wave modeling with high-order Boussinesq-type equations, *Journal of Waterway, Port, Coastal, and Ocean Engineering* **132**, 5, pp. 348–357, doi:https://doi.org/10.1061/(ASCE)0733-950X(2006)132:5(348).

Lynett, P. J., Gately, K., Wilson, R., Montoya, L., Arcas, D., Aytore, B., Bai, Y., Bricker, J. D., Castro, M. J., Cheung, K. F., *et al.* (2017). Inter-model analysis of tsunami-induced coastal currents, *Ocean Modelling* **114**, pp. 14–32, doi:https://doi.org/10.1016/j.ocemod.2017.04.003.

Lynett, P. J. and Liu, P. L.-F. (2004b). Linear analysis of the multi-layer model, *Coastal Engineering* **51**, pp. 439–454, doi:https://doi.org/10.1016/j.coastaleng.2004.05.004.

Ma, G., Shi, F., and Kirby, J. T. (2012). Shock-capturing non-hydrostatic model for fully dispersive surface wave processes, *Ocean Modelling* **43**, pp. 22–35, doi:https://doi.org/10.1016/j.ocemod.2011.12.002.

Ma, Q. W. (1998). Numerical simulation of nonlinear interaction between structures and steep waves, PhD thesis, University College London, UK.

Ma, Q., Wu, G. X., and Eatock-Taylor, R. (1997). Finite element analysis of non-linear transient waves in a three dimensional long tank, in *Proceedings of the 12th International Workshop on Water Waves and Floating Bodies*, Carry-le-Rouet, France.

Ma, Q. W., Wu, G. X., and Eatock Taylor, R. (2001a). Finite element simulation of fully non-linear interaction between vertical cylinders and steep waves.

Part 1: Methodology and numerical procedure, *International Journal for Numerical Methods in Fluids* **36**, 3, pp. 265–285, doi:https://doi.org/10. 1002/fld.131.

Ma, Q. W., Wu, G. X., and Eatock Taylor, R. (2001b). Finite element simulations of fully non-linear interaction between vertical cylinders and steep waves. Part 2: Numerical results and validation, *International Journal for Numerical Methods in Fluids* **36**, 3, pp. 287–308, doi:https://doi.org/10.1002/ fld.133.

Ma, Q. W. and Yan, S. (2006). Quasi ALE finite element method for nonlinear water waves, *Journal of Computational Physics* **212**, 1, pp. 52–72, doi: https://doi.org/10.1016/j.jcp.2005.06.014.

Ma, Q. W. and Yan, S. (2009). QALE-FEM for numerical modelling of non-linear interaction between 3D moored floating bodies and steep waves, *International Journal for Numerical Methods in Engineering* **78**, 6, pp. 713–756, doi:https://doi.org/10.1002/nme.2505.

Madsen, O. (1978). Wave-induced pore pressures and effective stresses in a porous bed, *Geotechnique* **28**, 4, pp. 377–393, doi:https://doi.org/10.1680/geot. 1978.28.4.377.

Madsen, O. S., Poon, Y.-K., and Graber, H. C. (1988). Spectral wave attenuation by bottom friction: Theory, in *Proceedings of the 21st International Conference on Coastal Engineering*.

Madsen, P., Murray, R., and Sørensen, O. R. (1992). A new form of the Boussinesq equations with improved linear dispersion characteristics. Part 2. A slowly-varying bathymetry, *Coastal Engineering* **18**, pp. 183–204, doi:https://doi. org/10.1016/0378-3839(92)90019-Q.

Madsen, P. A. and Agnon, Y. (2003). Accuracy and convergence of velocity formulations for water waves in the framework of Boussinesq theory, *Journal of Fluid Mechanics* **477**, pp. 285–319, doi:https://doi.org/10.1017/ S0022112002003257.

Madsen, P. A., Bingham, H. B., and Liu, H. (2002). A new Boussinesq method for fully nonlinear waves from shallow to deep water, *Journal of Fluid Mechanics* **462**, pp. 1–30, doi:https://doi.org/10.1017/S0022112002008467.

Madsen, P. A., Bingham, H. B., and Schäffer, H. A. (2003). Boussinesq-type formulations for fully nonlinear and extremely dispersive water waves: Derivation and analysis, *Proceedings of the Royal Society A: Maths, Physics, Engineering and Science* **459**, pp. 1075–1104, doi:https://doi.org/10.1098/ rspa.2002.1067.

Madsen, P. A. and Larsen, J. (1987). An efficient finite-difference approach to the mild-slope equation, *Coastal Engineering* **11**, 4, pp. 329–351, doi:https: //doi.org/10.1016/0378-3839(87)90032-9.

Madsen, P. A., Sgrensen, O., Schifer, H., Sørensen, O. R., Schäffer, H. A., Sgrensen, O., Schifer, H., Sørensen, O. R., and Schäffer, H. A. (1997). Surf zone dynamics simulated by a Boussinesq type model. Part I. Model description and cross-shore motion of regular waves, *Coastal Engineering* **32**, 4, pp. 255–287, doi:https://doi.org/10.1016/S0378-3839(97)00028-8.

Makin, V. and Kudryavtsev, V. (2002). Impact of dominant waves on sea drag, *Boundary-Layer Meteorology* **103**, 1, pp. 83–99, doi:https://doi.org/10.1023/A:1014591222717.

Makin, V., Kudryavtsev, V., and Mastenbroek, C. (1995). Drag of the sea surface, *Boundary-Layer Meteorology* **73**, 1, pp. 159–182, doi:https://doi.org/10.1007/BF00708935.

Makin, V. K. and Stam, M. (2003). *New Drag Formulation in NEDWAM* (KNMI, Koninklijk Nederlands Meteorologisch Instituut).

Malej, M., Shi, F., and Smith, J. M. (2019). Modeling ship-wake-induced sediment transport and morphological change–Sediment module in FUNWAVE-TVD. Tech. Rep. ERDC/CHL CHETN-VII-20, Coastal and Hydraulics Laboratory, Engineer Research and Development Center (US), doi:http://doi.org/10.21079/11681/32911.

Manasseh, R., Babanin, A. V., Forbes, C., Rickards, K., Bobevski, I., and Ooi, A. (2006). Passive acoustic determination of wave-breaking events and their severity across the spectrum, *Journal of Atmospheric and Oceanic Technology* **23**, 4, pp. 599–618, doi:https://doi.org/10.1175/JTECH1853.1.

Mancini, G., Briganti, R., McCall, R., Dodd, N., and Zhu, F. (2021). Numerical modelling of intra-wave sediment transport on sandy beaches using a non-hydrostatic, wave-resolving model, *Ocean Dynamics* **71**, 1, pp. 1–20, doi:https://doi.org/10.1007/s10236-020-01416-x.

Márquez, S. (2013). An extended mixture model for the simultaneous treatment of short and long scale interfaces, PhD thesis, Universidad Nacional del Litoral, Chile.

Marshall, J., Hill, C., Perelman, L., and Adcroft, A. (1997). Hydrostatic, quasi-hydrostatic, and nonhydrostatic ocean modeling, *Journal of Geophysical Research: Oceans* **102**, C3, pp. 5733–5752, doi:https://doi.org/10.1029/96JC02776.

Mase, H., Oki, K., Hedges, T. S., and Li, H. J. (2005). Extended energy-balance-equation wave model for multidirectional random wave transformation, *Ocean Engineering* **32**, 8–9, pp. 961–985, doi:https://doi.org/10.1016/j.oceaneng.2004.10.015.

Massel, S. R. (1993). Extended refraction-diffraction equation for surface waves, *Coastal Engineering* **19**, 1–2, pp. 97–126, doi:https://doi.org/10.1016/0378-3839(93)90020-9.

Mattosinho, G. O., Maciel, G. F., Ferreira, F. O., Santos, J. A., and Fortes, C. J. E. M. (2021). *A Review of FUNWAVE Model Applications in the Propagation of Waves Generated by Vessels* (CRC Press, Florida, USA), pp. 421–428, doi:https://doi.org/10.1201/9781003216599.

Maza, M., Lara, J. L., and Losada, I. J. (2016). Solitary wave attenuation by vegetation patches, *Advances in Water Resources* **98**, pp. 159–172, doi:https://doi.org/10.1016/j.advwatres.2016.10.021.

McCabe, M. and Stansby, P. K. (2010). An investigation on the coupling of spectral energy modelling and Boussinesq type modelling in the nearshore, in *Proceedings of the 1st European IAHR Congress*.

McKee, S., Tomé, M. F., Cuminato, J. A., Castelo, A., and Ferreira, V. G. (2004). Recent advances in the marker and cell method, *Archives of Computational Methods in Engineering* **11**, 2, p. 107, doi:https://doi.org/10.1007/BF02905936.

Ménard, T., Tanguy, S., and Berlemont, A. (2007). Coupling level set/VOF/ghost fluid methods: Validation and application to 3D simulation of the primary break-up of a liquid jet, *International Journal of Multiphase Flow* **33**, 5, pp. 510–524, doi:https://doi.org/10.1016/j.ijmultiphaseflow.2006.11.001.

Miles, J. W. (1957). On the generation of surface waves by shear flows, *Journal of Fluid Mechanics* **3**, 2, pp. 185–204, doi:https://doi.org/10.1017/S0022112057000567.

Mittal, R. and Iaccarino, G. (2005). Immersed boundary methods, *Annual Review of Fluid Mechanics* **37**, pp. 239–261, doi:https://doi.org/10.1146/annurev.fluid.37.061903.175743.

Monaghan, J. J. (1992). Smoothed particle hydrodynamics, *Annual Review of Astronomy and Astrophysics* **30**, 1, pp. 543–574, doi:https://doi.org/10.1146/annurev.aa.30.090192.002551.

Monaghan, J. J. (1994). Simulating free surface flows with SPH, *Journal of Computational Physics* **110**, 2, pp. 399–406, doi:https://doi.org/10.1006/jcph.1994.1034.

Moreira, R. M. and Peregrine, D. H. (2012). Nonlinear interactions between deep-water waves and currents, *Journal of Fluid Mechanics* **691**, pp. 1—25, doi:https://doi.org/10.1017/jfm.2011.436.

Mostafa, A. M., Mizutani, N., and Iwata, K. (1999). Nonlinear wave, composite breakwater, and seabed dynamic interaction, *Journal of Waterway, Port, Coastal, and Ocean Engineering* **125**, 2, pp. 88–97, doi:https://doi.org/10.1061/(ASCE)0733-950X(1999)125:2(88).

Nemati, F., Grilli, S. T., Ioualalen, M., Boschetti, L., Larroque, C., and Trevisan, J. (2019). High-resolution coastal hazard assessment along the French Riviera from co-seismic tsunamis generated in the Ligurian fault system, *Natural Hazards* **96**, 2, pp. 553–586, doi:https://doi.org/10.1007/s11069-018-3555-x.

New, A., McIver, P., and Peregrine, D. (1985). Computations of overturning waves, *Journal of Fluid Mechanics* **150**, pp. 233–251, doi:https://doi.org/10.1017/S0022112085000118.

Newell, A. C., Nazarenko, S., and Biven, L. (2001). Wave turbulence and intermittency, *Physica D: Nonlinear Phenomena* **152**, pp. 520–550, doi:https://doi.org/10.1016/S0167-2789(01)00192-0.

Nicholls, D. P. (1998). Traveling water waves: Spectral continuation methods with parallel implementation, *Journal of Computational Physics* **143**, 1, pp. 224–240, doi:https://doi.org/10.1006/jcph.1998.5957.

Nichols, B. D., Hirt, C. W., and Hotchkiss, R. (1980). SOLA-VOF: A solution algorithm for transient fluid flow with multiple free boundaries, Tech. rep., Los Alamos Scientific Lab, USA.

Ning, Y., Liu, W., Sun, Z., Zhao, X., and Zhang, Y. (2019). Parametric study of solitary wave propagation and runup over fringing reefs based on a Boussinesq wave model, *Journal of Marine Science and Technology* **24**, 2, pp. 512–525, doi:https://doi.org/10.1007/s00773-018-0571-1.

Noh, W. F. (1963). CEL: A time-dependent, two-space-dimensional, coupled Eulerian–Lagrange code, Tech. rep., Lawrence Radiation Laboratory, University of California, Livermore, doi:https://www.osti.gov/servlets/purl/4621975.

Noh, W. F. and Woodward, P. (1976). SLIC (Simple Line Interface Calculation), in *Proceedings of the 5th International Conference on Numerical Methods in Fluid Dynamics*, June 28–July 2, 1976, Twente University, Enschede, pp. 330–340, doi:https://doi.org/10.1007/3-540-08004-X_336.

Nwogu, O. (1993). Alternative form of Boussinesq equations for nearshore wave propagation, *Journal of Waterway, Port, Coastal and Ocean Engineering* **119**, pp. 618–638, doi:https://doi.org/10.1061/(ASCE)0733-950X(1993)119:6(618).

Ohnaka, S., Watanabe, A., and Isobe, M. (1989). Numerical modeling of wave deformation with a current, in *Proceedings of the 21st International Conference on Coastal Engineering*, pp. 393–407, doi:https://doi.org/10.1061/9780872626874.028.

Okamoto, T. and Basco, D. R. (2006). The relative trough Froude number for initiation of wave breaking: Theory, experiments and numerical model confirmation, *Coastal Engineering* **53**, 8, pp. 675–690, doi:https://doi.org/10.1016/j.coastaleng.2006.02.001.

Oki, K. and Sakai, T. (2009). Coupling of phase-averaging model and phase-resolving model by using double connection boundaries, in *Proceedings of Coastal Engineering Conference 2008*, pp. 302–313, doi:https://doi.org/10.1142/9789814277426_0025.

Okumura, H. (2014). A study of CUDA/MPI parallel computations for CADMAS-SURF/3D, in *Proceedings of the 24th International Ocean and Polar Engineering Conference*.

Oliveira, F. S. B. F. (2000). Improvement on open boundaries on a time dependent numerical model of wave propagation in the nearshore region, *Ocean Engineering* **28**, 1, pp. 95–115, doi:https://doi.org/10.1016/S0029-8018(99)00060-8.

Olsson, E. and Kreiss, G. (2005). A conservative level set method for two phase flow, *Journal of Computational Physics* **210**, 1, pp. 225–246, doi:https://doi.org/10.1016/j.jcp.2005.04.007.

Osher, S. and Sethian, J. A. (1988). Fronts propagating with curvature-dependent speed: Algorithms based on Hamilton-Jacobi formulations, *Journal of Computational Physics* **79**, 1, pp. 12–49, doi:https://doi.org/10.1016/0021-9991(88)90002-2.

Ouda, M. and Toorman, E. A. (2019). Development of a new multiphase sediment transport model for free surface flows, *International Journal of Multiphase Flow* **117**, pp. 81–102, doi:https://doi.org/10.1016/j.ijmultiphaseflow.2019.04.023.

Pan, C.-H., Lin, B.-Y., and Mao, X.-Z. (2007). Case study: Numerical modeling of the tidal bore on the Qiantang River, China, *Journal of Hydraulic Engineering* **133**, 2, pp. 130–138, doi:https://doi.org/10.1061/(ASCE) 0733-9429(2007)133:2(130).

Pan, J. N., Hong, G. W., and Zuo, Q. H. (2001). An extended mild-slope equation (in Chinese). *Ocean Engineering* **19**, 01, pp. 24–31.

Panchang, V. G. and Demirbilek, Z. (2001). Simulation of waves in harbors using two-dimensional elliptic equation models, *Advances in Coastal and Ocean Engineering*, pp. 125–162 doi:https://doi.org/10.1142/9789812794574_0003.

Panchang, V. G., Pearce, B. R., Wei, G., and Cushman-Roisin, B. (1991). Solution of the mild-slope wave problem by iteration, *Applied Ocean Research* **13**, 4, pp. 187–199, doi:https://doi.org/10.1016/S0141-1187(05)80074-4.

Panda, N., Dawson, C., Zhang, Y., Kennedy, A. B., Westerink, J. J., and Donahue, A. S. (2014). Discontinuous Galerkin methods for solving Boussinesq–Green–Naghdi equations in resolving non-linear and dispersive surface water waves, *Journal of Computational Physics* **273**, pp. 572–588, doi:https://doi.org/10.1016/j.jcp.2014.05.035.

Paris, R. and Ulvrova, M. (2019). Tsunamis generated by subaqueous volcanic explosions in Taal Caldera Lake, Philippines, *Bulletin of Volcanology* **81**, 3, pp. 1–14, doi:https://doi.org/10.1007/s00445-019-1272-2.

Park, H., Do, T., Tomiczek, T., Cox, D. T., and van de Lindt, J. W. (2018). Numerical modeling of non-breaking, impulsive breaking, and broken wave interaction with elevated coastal structures: Laboratory validation and inter-model comparisons, *Ocean Engineering* **158**, pp. 78–98, doi:https://doi.org/10.1016/j.oceaneng.2018.03.088.

Passenko, J., Lessin, G., Erichsen, A. C., and Raudsepp, U. (2008). Validation of hydrostatic and non-hydrostatic versions of the hydrodynamical model MIKE 3 applied for the Baltic Sea, *Estonian Journal of Engineering* **14**, 3, doi:https://doi.org/10.3176/eng.2008.3.05.

Patera, A. T. (1984). A spectral element method for fluid dynamics: Laminar flow in a channel expansion, *Journal of Computational Physics* **54**, 3, pp. 468–488, doi:https://doi.org/10.1016/0021-9991(84)90128-1.

Pearce, B. R. and Panchang, V. G. (1985). A method for investigation of steady state wave spectra in bays, *Journal of Waterway, Port, Coastal and Ocean Engineering* **111**, 4, pp. 629–644, doi:https://doi.org/10.1061/(ASCE)0733-950X(1985)111:4(629).

Pedlosky, J. (1987). *Geophysical Fluid Dynamics* (Springer-Verlag, New York), doi:https://doi.org/10.1007/978-1-4612-4650-3.

Peregrine, D. H. (1967). Long waves on a beach, *Journal of Fluid Mechanics* **27**, pp. 815–827, doi:https://doi.org/10.1017/S0022112067002605.

Perumal, D. A. and Dass, A. K. (2015). A review on the development of lattice Boltzmann computation of macro fluid flows and heat transfer, *Alexandria Engineering Journal* **54**, 4, pp. 955–971, doi:https://doi.org/10.1016/j.aej.2015.07.015.

Peskin, C. S. (2002). The immersed boundary method, *Acta Numerica* **11**, pp. 479–517, doi:https://doi.org/10.1017/S0962492902000077.

Phillips, O. (1985). Spectral and statistical properties of the equilibrium range in wind-generated gravity waves, *Journal of Fluid Mechanics* **156**, pp. 505–531, doi:https://doi.org/10.1017/S0022112085002221.

Phillips, O. M. (1957). On the generation of waves by turbulent wind, *Journal of Fluid Mechanics* **2**, 5, pp. 417–445, doi:https://doi.org/10.1017/S0022112057000233.

Phillips, O. M. and Hasselmann, K. (1986). *Wave Dynamics and Radio Probing of the Ocean Surface* (Springer Science & Business Media, Springer, New York, USA), doi:https://doi.org/10.1007/978-1-4684-8980-4.

Phuoc, V. L. and Massel, S. (2008). Energy dissipation in non-uniform mangrove forests of arbitrary depth, *Journal of Marine Systems* **74**, 1–2, pp. 603–622, doi:https://doi.org/10.1016/j.jmarsys.2008.05.004.

Polnikov, V. (1993). On a description of a wind-wave energy dissipation function, in *Proceedings of the Air–Sea Interface Symposium: Radio and Acoustic Sensing, Turbulence and Wave Dynamics*, Vol. 277 (Rosenstiel School of Marine and Atmospheric Science, University of Miami, USA), p. 282.

Ponce de Leon, S. and Osborne, A. R. (2020). Role of nonlinear four-wave interactions source term on the spectral shape, *Journal of Marine Science and Engineering* **8**, 4, doi:https://doi.org/10.3390/jmse8040251.

Popinet, S. (2018). Numerical models of surface tension, *Annual Review of Fluid Mechanics* **50**, pp. 49–75, doi:https://doi.org/10.1146/annurev-fluid-122316-045034.

Porter, D. (2003). The mild-slope equations, *Journal of Fluid Mechanics* **494**, doi:https://doi.org/10.1017/S0022112003005846.

Qi, J., Chen, C., Beardsley, R. C., Perrie, W., Cowles, G. W., and Lai, Z. (2009). An unstructured-grid finite-volume surface wave model (FVCOM-SWAVE): Implementation, validations and applications, *Ocean Modelling* **28**, 1, pp. 153–166, doi:https://doi.org/10.1016/j.ocemod.2009.01.007.

Quinn, B., Toledo, Y., and Shrira, V. (2017). Explicit wave action conservation for water waves on vertically sheared flows, *Ocean Modelling* **112**, pp. 33–47, doi:https://doi.org/10.1016/j.ocemod.2017.03.003.

Radder, A. C. (1979). On the parabolic equation method for water-wave propagation, *Journal of Fluid Mechanics* **95**, 1, pp. 159–176, doi:https://doi.org/10.1017/S0022112079001397.

Ransley, E., Yan, S., Brown, S. A., Mai, T., Graham, D., Ma, Q., Musiedlak, P.-H., Engsig-Karup, A. P., Eskilsson, C., Li, Q., *et al.* (2019). A blind comparative study of focused wave interactions with a fixed FPSO-like structure (CCP-WSI Blind Test Series 1), *International Journal of Offshore and Polar Engineering* **29**, 02, pp. 113–127, doi:https://doi.org/10.17736/ijope.2019.jc748.

Ransley, E. J., Brown, S. A., Hann, M., Greaves, D. M., Windt, C., Ringwood, J., Davidson, J., Schmitt, P., Yan, S., Wang, J., Wang, J., Ma, Q., Xie, Z., Giorgi, G., Hughes, J., Williams, A., Masters, I., Lin, Z., Chen, H., Qian, L., Ma, Z., Chen, Q., Ding, H., Zang, J., van Rij, J., Yu, Y.-H., Li, Z., Bouscasse, B., Ducrozet, G., and Bingham, H. (2021). Focused wave interactions with floating structures: A blind comparative study, *Proceedings of the Institution of Civil Engineers - Engineering and Computational Mechanics* **174**, 1, pp. 46–61, doi:https://doi.org/10.1680/jencm.20.00006.

Reniers, A. and Zijlema, M. (2022). SWAN SurfBeat-1D, *Coastal Engineering* **172**, doi:https://doi.org/10.1016/j.coastaleng.2021.104068.

Reynolds, O. (1895). On the dynamical theory of incompressible viscous fluids and the determination of the criterion, *Philosophical Transactions of the Royal Society of London (A)* **186**, pp. 123–164, doi:https://doi.org/10.1098/rsta. 1895.0004.

Rijnsdorp, D. P., Hansen, J. E., and Lowe, R. J. (2018). Simulating the wave-induced response of a submerged wave-energy converter using a non-hydrostatic wave-flow model, *Coastal Engineering* **140**, pp. 189–204, doi: https://doi.org/10.1016/j.coastaleng.2018.07.004.

Rijnsdorp, D. P., Reniers, A. J., and Zijlema, M. (2021). Free infragravity waves in the north sea, *Journal of Geophysical Research: Oceans* **126**, 8, p. e2021JC017368, doi:https://doi.org/10.1029/2021JC017368.

Rijnsdorp, D. P. and Zijlema, M. (2016). Simulating waves and their interactions with a restrained ship using a non-hydrostatic wave-flow model. *Coastal Engineering* **114**, pp. 119–136, doi:https://doi.org/10.1016/j.coastaleng. 2016.04.018.

Risandi, J., Rijnsdorp, D. P., Hansen, J. E., and Lowe, R. J. (2020). Hydrodynamic modeling of a reef-fringed pocket beach using a phase-resolved non-hydrostatic model, *Journal of Marine Science and Engineering* **8**, 11, p. 877, doi:https://doi.org/10.3390/jmse8110877.

Robertson, I. and Sherwin, S. (1999). Free-surface flow simulation using hp/spectral elements, *Journal of Computational Physics* **155**, 1, pp. 26–53, doi:https://doi.org/10.1006/jcph.1999.6328.

Roe, P. L. (1981). Approximate Riemann solvers, parameter vectors, and difference schemes, *Journal of Computational Physics* **43**, 2, pp. 357–372, doi:https://doi.org/10.1016/0021-9991(81)90128-5.

Roeber, V. and Cheung, K. F. (2012). Boussinesq-type model for energetic breaking waves in fringing reef environments, *Coastal Engineering* **70**, pp. 1–20, doi:https://doi.org/10.1016/j.coastaleng.2012.06.001.

Roenby, J., Bredmose, H., and Jasak, H. (2016). A computational method for sharp interface advection. *Royal Society Open Science* **3**, 11, doi:https:// doi.org/10.1098/rsos.160405.

Rogers, W. E., Hwang, P. A., and Wang, D. W. (2003). Investigation of wave growth and decay in the SWAN model: Three regional-scale applications, *Journal of Physical Oceanography* **33**, 2, pp. 366–389, doi:https://doi.org/ 10.1175/1520-0485(2003)033⟨0366:IOWGAD⟩2.0.CO;2.

Roland, A. (2008). Development of WWM II: Spectral wave modelling on unstructured meshes, PhD thesis, Technische Universität Darmstadt, Institute of Hydraulic and Water Resources Engineering, Germany.

Rutten, J., Torres-Freyermuth, A., and Puleo, J. A. (2021). Uncertainty in runup predictions on natural beaches using XBeach nonhydrostatic, *Coastal Engineering* **166**, doi:https://doi.org/10.1016/j.coastaleng.2021.103869.

Ryu, S., Kim, M., and Lynett, P. (2003). Fully nonlinear wave-current interactions and kinematics by a BEM-based numerical wave tank, *Computational Mechanics* **32**, pp. 336–346, doi:https://doi.org/10.1007/ s00466-003 0491-7.

Salas Pérez, M. (2014). Overtopping over a real rubble mound breakwater calculated with SWASH, Bachelor thesis, Delft University of Technology, Delft.

Sasikumar, A., Kamath, A., Musch, O., Bihs, H., and Arntsen, Ø. A. (2019). Numerical modeling of berm breakwater optimization with varying berm geometry using REEF3D, *Journal of Offshore Mechanics and Arctic Engineering* **141**, 1, doi:https://doi.org/10.1115/1.4040508.

Schäffer, H. A. (2008). Comparison of Dirichlet–Neumann operator expansions for nonlinear surface gravity waves, *Coastal Engineering* **55**, 4, pp. 288–294, doi:https://doi.org/10.1016/j.coastaleng.2007.11.002.

Schäffer, H. A. and Madsen, P. A. (1995). Further enhancements of Boussinesq-type equations, *Coastal Engineering* **26**, pp. 1–14, doi:https://doi.org/10.1016/0378-3839(95)00017-2.

Schäffer, H. A., Madsen, P. A., and Deigaard, R. (1993). A Boussinesq model for waves breaking in shallow water, *Coastal Engineering* **20**, 3–4, pp. 185–202, doi:https://doi.org/10.1016/0378-3839(93)90001-O.

Serre, F. (1953). Contribution à l'étude des écoulements permanents et variables dans les canaux (in French – Contribution to the study of permanent and non-permanent flows in channels), *La Houille Blanche – The Wave* **6**, pp. 830–872, doi:https://doi.org/10.1051/lhb/1953034.

Shan, X. and Chen, H. (1993). Lattice Boltzmann model for simulating flows with multiple phases and components, *Physical Review E* **47**, 3, p. 1815, doi:https://doi.org/10.1103/PhysRevE.47.1815.

Shankar, J., Cheong, H.-F., and Chan, C.-T. (1997). Boundary fitted grid models for tidal motions in Singapore coastal waters, *Journal of Hydraulic Research* **35**, 1, pp. 3–20, doi:https://doi.org/10.1080/00221689709498641.

Shao, S. (2006). Incompressible SPH simulation of wave breaking and overtopping with turbulence modelling, *International Journal for Numerical Methods in Fluids* **50**, pp. 597–621, doi:https://doi.org/10.1002/fld.1068.

Sharma, A., Panchang, V. G., and Kaihatu., J. M. (2014). Modeling nonlinear wave–wave interactions with the elliptic mild slope equation, *Applied Ocean Research* **48**, pp. 114–125, doi:https://doi.org/10.1016/j.apor.2014.08.004.

Shchepetkin, A. F. and McWilliams, J. C. (2003). A method for computing horizontal pressure-gradient force in an oceanic model with a nonaligned vertical coordinate, *Journal of Geophysical Research: Oceans* **108**, C3, doi: https://doi.org/10.1029/2001JC001047.

Shelby, M., Grilli, S. T., and Grilli, A. R. (2016). Tsunami hazard assessment in the Hudson River Estuary based on dynamic tsunami–tide simulations, in E. L. Geist, H. M. Fritz, A. B. Rabinovich, Y.Tanioka (ed.), *Global Tsunami Science: Past and Future, Volume I* (Springer, Springer, Berlin, Germany), pp. 3999–4037, doi:https://doi.org/10.1007/978-3-319-55480-8_17.

Shen, L. and Chan, E.-S. (2008). Numerical simulation of fluid-structure interaction using a combined volume of fluid and immersed boundary method, *Ocean Engineering* **35**, 8–9, pp. 939–952, doi:https://doi.org/10.1016/j.oceaneng.2008.01.013.

Shen, Z., Wan, D., and Carrica, P. M. (2015). Dynamic overset grids in Open-FOAM with application to KCS self-propulsion and maneuvering, *Ocean*

Engineering **108**, pp. 287–306, doi:https://doi.org/10.1016/j.oceaneng.2015.07.035.

Shi, F., Dalrymple, R. A., Kirby, J. T., Chen, Q., and Kennedy, A. (2001). A fully nonlinear Boussinesq model in generalized curvilinear coordinates, *Coastal Engineering* **42**, 4, pp. 337–358, doi:https://doi.org/10.1016/S0378-3839(00)00067-3.

Shi, F., Kirby, J. T., Harris, J. C., Geiman, J. D., and Grilli, S. T. (2012). A high-order adaptive time-stepping TVD solver for Boussinesq modeling of breaking waves and coastal inundation, *Ocean Modelling* **43–44**, pp. 36–51, doi:https://doi.org/10.1016/j.ocemod.2011.12.004.

Shi, F., Ma, G., and Kirby, J. T. (2010). Numerical modeling of optical properties inside the surfzone, in *Proceedings of the 32nd International Conference on Coastal Engineering (ICCE)*, Shanghai, doi:https://doi.org/10.9753/icce.v32.currents.51.

Shields, J. J. and Webster, W. C. (1988). On direct methods in water wave theory, *Journal of Fluid Mechanics* **197**, pp. 171–199, doi:https://doi.org/10.1017/S0022112088003222.

Shuto, N., Goto, C., and Imamura, F. (1990). Numerical simulation as a means of warning for near-field tsunamis, *Coastal Engineering in Japan* **33**, 2, pp. 173–193, doi:https://doi.org/10.1080/05785634.1990.11924532.

Sitanggang, K. I. and Lynett, P. J. (2010). Multi-scale simulation with a hybrid Boussinesq-RANS hydrodynamic model, *International Journal for Numerical Methods in Fluids* **62**, 9, pp. 1013–1046, doi:https://doi.org/10.1002/fld.2056.

Sitanggang, K. I., Lynett, P. J., and Liu, P. L.-F. (2007). Development of a Boussinesq-RANS VOF hybrid wave model, in *Coastal Engineering 2006*, pp. 24–35, doi:https://doi.org/10.1142/9789812709554_0003.

Slattery, J. C. (1967). Flow of viscoelastic fluids through porous media, *American Institute of Chemical Engineers Journal* **13**, 6, pp. 1066–1071, doi:https://doi.org/10.1002/aic.690130606.

Smagorinsky, J., Manabe, S., and Holloway Jr, J. L. (1965). Numerical results from a nine-level general circulation model of the atmosphere, *Monthly Weather Review* **93**, 12, pp. 727–768, doi:https://doi.org/10.1175/1520-0493(1965)093⟨0727:NRFANL⟩2.3.CO;2.

Smit, P. B., Stelling, G. S., Roelvink, D., van Thiel de Vries, J., McCall, R., van Dongeren, A., Zwinkels, C., and Jacobs, R. (2010). XBeach: Non-hydrostatic model. Validation, verification and model description, Tech. rep., Delft University of Technology and Deltares, Netherland, doi:https://oss.deltares.nl/documents/48999/49476/non-hydrostatic_report_draft.pdf.

Smith, R. and Sprinks, T. (1975). Scattering of surface waves by a conical island, *Journal of Fluid Mechanics* **72**, 2, pp. 373–384, doi:https://doi.org/10.1017/S0022112075003424.

Smith, R. A. (1998). An operator expansion formalism for nonlinear surface waves over variable depth, *Journal of Fluid Mechanics* **363**, pp. 333–347, doi:https://doi.org/10.1017/S0022112098001219.

Song, C.-G., Ryu, Y., and Jung, T.-H. (2015). A new model for wave transformation around axis-symmetric islands without vertical wall along the coastline, *Applied Ocean Research* **51**, pp. 67–73, doi:https://doi.org/10.1016/j.apor.2015.02.010.

Song, Z., Zhang, H., Kong, J., Li, R., and Zhang, W. (2007). An efficient numerical model of hyperbolic mild-slope equation, in *Proceedings of the 26th International Conference on Offshore Mechanics and Arctic Engineering*, pp. 253–258, doi:https://doi.org/10.1115/OMAE2007-29146.

Souli, M. and Zolesio, J. P. (2001). Arbitrary Lagrangian–Eulerian and free surface methods in fluid mechanics, *Computer Methods in Applied Mechanics and Engineering* **191**, 3–5, pp. 451–466, doi:https://doi.org/10.1016/S0045-7825(01)00313-9.

Spalart, P. and Allmaras, S. (1992). A one-equation turbulence model for aerodynamic flows, in *30th Aerospace Sciences Meeting and Exhibit*.

Sriram, V., Sannasiraj, S., and Sundar, V. (2006). Simulation of 2-D nonlinear waves using finite element method with cubic spline approximation, *Journal of Fluids and Structures* **22**, 5, pp. 663–681, doi:https://doi.org/10.1016/j.jfluidstructs.2006.02.007.

Stelling, G. and Zijlema, M. (2003a). An accurate and efficient finite-difference algorithm for non-hydrostatic free-surface flow with application to wave propagation, *International Journal for Numerical Methods in Fluids* **43**, 1, pp. 1–23, doi:https://doi.org/10.1002/fld.595.

Stelling, G. and Zijlema, M. (2003b). An accurate and efficient finite-difference algorithm for non-hydrostatic free-surface flow with application to wave propagation, *International Journal for Numerical Methods in Fluids* **23**, pp. 1–23, doi:https://doi.org/10.1002/fld.595.

Stelling, G. S. and Duinmeijer, S. A. (2003). A staggered conservative scheme for every froude number in rapidly varied shallow water flows, *International Journal for Numerical Methods in Fluids* **43**, 12, pp. 1329–1354, doi:https://doi.org/10.1002/fld.537.

Stive, M. J. F. (1985). A scale comparison of waves breaking on a beach, *Coastal Engineering* **9**, 2, pp. 151–158, doi:https://doi.org/10.1016/0378-3839(85)90003-1.

Su, C. H. and Gardner, C. S. (1969). Korteweg-de Vries equation and generalizations. III. Derivation of the Korteweg-de Vries equation and Burgers equation, *Journal of Mathematical Physics* **10**, pp. 536–539, doi:https://doi.org/10.1063/1.1664873.

Sudhakar, T. and Das, A. K. (2020). Evolution of multiphase Lattice Boltzmann method: A review, *Journal of The Institution of Engineers (India): Volume 101*, pp. 1–9, doi:https://doi.org/10.1007/s40032-020-00600-8.

Suh, K. D., Lee, C., and Park, W. S. (1997). Time-dependent equations for wave propagation on rapidly varying topography, *Coastal Engineering* **32**, 2–3, pp. 91–117, doi:https://doi.org/10.1016/S0378-3839(97)81745-0.

Sumer, B. M. and Fuhrman, D. R. (2020). *Turbulence in Coastal and Civil Engineering*. Vol. 51 (World Scientific, Singapore).

Sun, R. and Xiao, H. (2016). SediFoam: A general-purpose, open-source CFD–DEM solver for particle-laden flow with emphasis on sediment transport, *Computers & Geosciences* **89**, pp. 207–219, doi:https://doi.org/10.1016/j.cageo.2016.01.011.

Sung, H. G. and Grilli, S. T. (2005). Numerical modeling of nonlinear surface waves caused by surface effect ships dynamics and kinematics, in *Proceedings of the 15th International Ocean and Polar Engineering Conference*, pp. 124–131.

Sung, H. G. and Grilli, S. T. (2006a). BEM Computations of 3-D fully nonlinear free surface flows caused by advancing surface disturbances, *International Journal of Offshore and Polar Engineering* **18**, 04, pp. 292–301.

Sung, H. G. and Grilli, S. T. (2006b). Combined Eulerian–Lagrangian or pseudo-Lagrangian descriptions of waves caused by an advancing free surface disturbance, in *Proceedings of the 16th International Ocean and Polar Engineering Conference*, pp. 487–494.

Suzuki, T., Verwaest, T., Veale, W., Trouw, K., and Zijlema, M. (2012). A numerical study on the effect of beach nourishment on wave overtopping in shallow foreshores, in *Proceedings of the 33rd International Conference on Coastal Engineering, ICCE 2012*, Santander, doi:https://doi.org/10.9753/icce.v33.waves.50.

SWAN Team (2020). Swan scientific and technical documentation, Tech. rep., Delft University of Technology, Delft, Holland, doi:http://swanmodel.sourceforge.net/download/zip/swantech.pdf.

SWASH Team (2020). Swash user manual, Tech. rep., Delft University of Technology, Delft, Holland, doi:http://swash.sourceforge.net/download/zip/swashuse.pdf.

Swift, M. R., Orlandini, E., Osborn, W. R., and Yeomans, J. M. (1996). Lattice Boltzmann simulations of liquid-gas and binary fluid systems, *Physical Review E* **54**, 5, p. 5041, doi:https://doi.org/10.1103/PhysRevE.54.5041.

Tanaka, M. (1983). The stability of steep gravity waves, *Journal of the Physical Society of Japan* **52**, 9, pp. 3047–3055, doi:https://doi.org/10.1143/JPSJ.52.3047.

Tanaka, M. (1985). The stability of steep gravity waves. Part 2, *Journal of Fluid Mechanics* **156**, pp. 281–289, doi:https://doi.org/10.1017/S0022112085002099.

Tanaka, M. (1986). The stability of solitary waves, *The Physics of Fluids* **29**, 3, pp. 650–655, doi:https://doi.org/10.1063/1.865459.

Tang, J., Shen, S., and Wang., H. (2015). Numerical model for coastal wave propagation through mild slope zone in the presence of rigid vegetation, *Coastal Engineering* **97**, pp. 53–59, doi:https://doi.org/10.1016/j.coastaleng.2014.12.006.

Tang, J., Shen, Y., Zheng, Y., and Qiu, D. (2004). An efficient and flexible computational model for solving the mild slope equation, *Coastal Engineering* **51**, 2, pp. 143–154, doi:https://doi.org/10.1016/j.coastaleng.2003.12.005.

Tao, J. (1983a). Computation of wave runup and wave breaking, Internal Report, Danish Hydraulics Institute, Denmark.

Tao, J. (1983b). Numerical modeling of wave runup and breaking on the beach, *Acta Oceanologica Sinica* **6**, pp. 692–700.

Tao, J. H. and Han, G. (2001). Numerical simulation of breaking wave based on higher-order mild slope equation, *China Ocean Engineering* **15**, 2, pp. 269–280.

Tappin, D. R., Watts, P., and Grilli, S. T. (2008). The Papua New Guinea tsunami of 17 July 1998: Anatomy of a catastrophic event, *Natural Hazards and Earth System Sciences* **8**, 2, pp. 243–266, doi:https://doi.org/10.5194/nhess-8-243-2008.

Tehranirad, B., Kirby, J. T., Grilli, S., and Shi, F. (2017). Does a morphological adjustment during tsunami inundation increase levels of hazards? in *Proceedings of the Coastal Structures and Solutions to Coastal Disasters 2015: Tsunamis*, pp. 145–153, doi:https://doi.org/10.1061/9780784480311.015.

Temam, R. and Ziane, M. (2005). Chap. 6 Some mathematical problems in geophysical fluid dynamics, in S. Friedlander and D. Serre (eds.), *Handbook of Mathematical Fluid Dynamics*, Vol. 3 (North-Holland, Elsevier, Amsterdam, NL), pp. 535–658, doi:https://doi.org/10.1016/S1874-5792(05)80009-6.

Thorimbert, Y., Latt, J., Cappietti, L., and Chopard, B. (2016). Virtual wave flume and oscillating water column modeled by lattice Boltzmann method and comparison with experimental data, *International Journal of Marine Energy* **14**, pp. 41–51, doi:https://doi.org/10.1016/j.ijome.2016.04.001.

Thorimbert, Y., Lätt, J., and Chopard, B. (2019). Coupling of lattice Boltzmann shallow water model with lattice Boltzmann free-surface model, *Journal of Computational Science* **33**, pp. 1–10, doi:https://doi.org/10.1016/j.jocs.2019.01.006.

Thornton, E. B. and Guza, R. T. (1983). Transformation of wave height distribution, *Journal of Geophysical Research: Oceans* **88**, C10, pp. 5925–5938, doi:https://doi.org/10.1029/JC088iC10p05925.

Titov, V. V. and Synolakis, C. E. (1998). Numerical modeling of tidal wave runup, *Journal of Waterway, Port, Coastal, and Ocean Engineering* **124**, 4, pp. 157–171, doi:https://doi.org/10.1061/(ASCE)0733-950X(1998)124:4(157).

Toledo, Y., Hsu, T. W., and Roland, A. (2012). Extended time-dependent mildslope and wave-action equations for wave–bottom and wave–current interactions, *Proceedings of the Royal Society A: Mathematical, Physical and Engineering Sciences* **468**, 2137, pp. 184–205, doi:https://doi.org/10.1098/rspa.2011.0377.

Tolman, H. L. (1990). The influence of unsteady depths and currents of tides on wind-wave propagation in shelf seas, *Journal of Physical Oceanography* **20**, 8, pp. 1166–1174, doi:https://doi.org/10.1175/1520-0485(1990)020⟨1166:TIOUDA⟩2.0.CO;2.

Tolman, H. L. (1991). A third-generation model for wind waves on slowly varying, unsteady and inhomogeneous depths and currents, *Journal of Physical Oceanography* **21**, 6, pp. 782–797, doi:https://doi.org/10.1175/1520-0485(1991)021⟨0782:ATGMFW⟩2.0.CO;2.

Tolman, H. L. (1994). Wind waves and moveable-bed bottom friction, *Journal of Physical Oceanography* **24**, 5, pp. 994–1009, doi:https://doi.org/10.1175/1520-0485(1994)024⟨0994:WWAMBB⟩2.0.CO;2.

Tolman, H. L. (2009). User manual and system documentation of WAVEWATCH III TM version 3.14, Technical note, MMAB Contribution **276**, p. 220.

Tom, J. G., Rijnsdorp, D. P., Ragni, R., and White, D. J. (2019). Fluid-structure-soil interaction of a moored wave energy device, in *Proceedings of the 38th International Conference on Offshore Mechanics and Arctic Engineering*, doi:https://doi.org/10.1115/OMAE2019-95419.

Tonelli, M. and Petti, M. (2009). Hybrid finite volume–finite difference scheme for 2DH improved Boussinesq equations, *Coastal Engineering* **56**, 5–6, pp. 609–620, doi:https://doi.org/10.1016/j.coastaleng.2009.01.001.

Tong, F. F., Shen, Y. M., Jun, T., and Lei, C. (2010). Water wave simulation in curvilinear coordinates using a time-dependent mild slope equation, *Journal of Hydrodynamics, Series B* **22**, 6, pp. 796–803, doi:https://doi.org/10.1016/S1001-6058(09)60118-9.

Tong, R. P. (1997). A new approach to modelling an unsteady free surface in boundary integral methods with application to bubble-structure interactions, *Mathematics and Computers in Simulation* **44**, 4, pp. 415–426, doi:https://doi.org/10.1016/S0378-4754(97)00067-0.

Toro, E. F. (2013). *Riemann Solvers and Numerical Methods for Fluid Dynamics: A Practical Introduction* (Springer Science & Business Media, NY, USA).

Torres-Freyermuth, A., Lara, J. L., and Losada, I. J. (2010). Numerical modelling of short- and long-wave transformation on a barred beach. *Coastal Engineering* **57**, pp. 317–330, doi:https://doi.org/10.1016/j.coastaleng.2009.10.013.

Touboul, J. (2007). On the influence of wind on extreme wave events, *Natural Hazards and Earth System Sciences* **7**, 1, pp. 123–128, doi:https://doi.org/10.5194/nhess-7-123-2007.

Touboul, J., Charland, J., Rey, V., and Belibassakis, K. (2016). Extended mild-slope equation for surface waves interacting with a vertically sheared current, *Coastal Engineering* **116**, pp. 77–88, doi:https://doi.org/10.1016/j.coastaleng.2016.06.003.

Touboul, J. and Kharif, C. (2010). Two-dimensional direct numerical simulations of the dynamics of rogue waves under wind action. chap. 2, Q. Ma (ed.), *Advances in Numerical Simulation of Nonlinear Water Waves* (World Scientific, Singapore), pp. 43–74, doi:https://doi.org/10.1142/9789812836502_0002.

Touboul, J. and Kharif, C. (2016). Effect of vorticity on the generation of rogue waves due to dispersive focusing, *Natural Hazards* **84**, pp. 585–598, doi:https://doi.org/10.1007/s11069-016-2419-5.

Troch, P. and De Rouck, J. (1998). Development of 2D numerical wave flume for simulation of wave interaction with rubble mound breakwaters, *Proceedings 26th International Conference on Coastal Engineering (ICCE)*, Copenhagen. doi:https://doi.org/10.1061/9780784404119.122.

Tsai, C. P., Chen, H. B., and Hsu, J. R. C. (2001). Calculations of wave transformation across the surf zone, *Ocean Engineering* **28**, 8, pp. 941–955, doi: https://doi.org/10.1016/S0029-8018(00)00047-0.

Tsai, C. P., Chen, H. B., and Lee, F. C. (2006). Wave transformation over submerged permeable breakwater on porous bottom, *Ocean Engineering* **33**, 11–12, pp. 1623–1643, doi:https://doi.org/10.1016/j.oceaneng.2005.09.006.

Tuck, E. O. and Hwang, L. S. (1972). Long wave generation on a sloping beach. *Journal of Fluid Mechanics* **51**, 3, pp. 449–461, doi:https://doi.org/10.1017/S0022112072002289.

Turnbull, M., Borthwick, A., and Eatock-Taylor, R. (2003). Wave–structure interaction using coupled structured–unstructured finite element meshes, *Applied Ocean Research* **25**, 2, pp. 63–77, doi:https://doi.org/10.1016/S0141-1187(03)00032-4.

Ullmann, S. (2008). *Three-dimensional computation of non-hydrostatic free-surface flows*, Master's thesis, Delft University of Technology, Delft.

Vachaparambil, K. J. and Einarsrud, K. E. (2019). Comparison of surface tension models for the volume of fluid method, *Processes* **7**, 8, doi:https://doi.org/10.3390/pr7080542.

van der Westhuysen, A. J., Zijlema, M., and Battjes, J. A. (2007). Nonlinear saturation-based whitecapping dissipation in SWAN for deep and shallow water, *Coastal Engineering* **54**, 2, pp. 151–170, doi:https://doi.org/10.1016/j.coastaleng.2006.08.006.

van Gent, M. R. A. (1995). Porous flow through rubble-mound material, *Journal of Waterway, Port, Coastal and Ocean Engineering* **3**, 121, pp. 176–181, doi:https://doi.org/10.1061/(ASCE)0733-950X(1995)121:3(176).

Van Ormondt, M., Nederhoff, K., and van Dongeren, A. (2020). Delft Dashboard: a quick set-up tool for hydrodynamic models, *Journal of Hydroinformatics* **22**, 3, pp. 510–527, doi:https://doi.org/10.2166/hydro.2020.092.

Van Vledder, P. G. (2001). Improved algorithms for computing the non-linear quadruplet wave-wave interactions in deep and shallow water, in *ECMWF Workshop on Ocean Wave Forecasting, European Centre for Medium-range Weather Forecasts Reading*, England, United Kingdom, doi:https://www.ecmwf.int/sites/default/files/elibrary/2001/13251-improved-algorithms-computing-non-linear-quadruplet-wave-wave-interactions-deep-and-shallow.pdf.

Veeramony, J., Condon, A., and van Ormondt, M. (2017). Forecasting storm surge and inundation: Model validation, *Weather and Forecasting* **32**, 6, pp. 2045–2063, doi:https://doi.org/10.1175/WAF-D-17-0015.1.

Veeramony, J., Orzech, M. D., Edwards, K. L., Gilligan, M., Choi, J., Terrill, E., and De Paolo, T. (2014). Navy nearshore ocean prediction systems, *Oceanography* **27**, 3, pp. 80–91, doi:https://www.jstor.org/stable/24862191.

Veeramony, J. and Svendsen, I. A. (2000). The flow in surf-zone waves, *Coastal Engineering* **39**, pp. 93–122, doi:https://doi.org/10.1016/S0378-3839(99)00058-7.

Vinje, T. and Brevig, P. (1981). Numerical simulation of breaking waves, *Advances in Water* **4**, 2, pp. 77–82, doi:https://doi.org/10.1016/0309-1708(81) 90027-0.

Vukcevic, V., Jasak, H., and Gatin, I. (2017). Implementation of the ghost fluid method for free surface flows in polyhedral finite volume framework, *Computers & Fluids* **153**, pp. 1–19, doi:https://doi.org/10.1016/j.compfluid. 2017.05.003.

Vyzikas, T., Stagonas, D., Buldakov, E., Greaves, D., (2015). Efficient numerical modelling of focused wave groups for freak wave generation, in *Proceedings of the 25th International Ocean and Polar Engineering Conference*, doi: https://www.onepetro.org/conference-paper/ISOPE-I-15-693.

Wacławczyk, T. and Koronowicz, T. (2008). Comparison of CICSAM and HRIC high-resolution schemes for interface capturing, *Journal of Theoretical and Applied Mechanics* **46**, 2, pp. 325–345.

Wallcraft, A. J., Metzger, E. J., and Carroll, S. N. (2009). Software design description for the hybrid coordinate ocean model (HYCOM), Version 2.2, Tech. rep., Naval Research Lab, Stennis Space Center (MS), Oceanography division, USA.

WAMDI Team (1988). The WAM model - A third generation ocean wave prediction model, *Journal of Physical Oceanography* **18**, 12, pp. 1775–1810, doi: https://doi.org/10.1175/1520-0485(1988)018⟨1775:TWMTGO⟩2.0.CO;2.

Wang, C. Z. and Khoo, B. C. (2005). Finite element analysis of two-dimensional nonlinear sloshing problems in random excitations, *Ocean Engineering* **32**, 2, pp. 107–133, doi:https://doi.org/10.1016/j.oceaneng.2004.08.001.

Wang, C. Z. and Wu, G. X. (2006). An unstructured-mesh-based finite element simulation of wave interactions with non-wall-sided bodies, *Journal of Fluids and Structures* **22**, 4, pp. 441–461, doi:https://doi.org/10.1016/j. jfluidstructs.2005.12.005.

Wang, C. Z. and Wu, G. X. (2007). Time domain analysis of second-order wave diffraction by an array of vertical cylinders, *Journal of Fluids and Structures* **23**, 4, pp. 605–631, doi:https://doi.org/10.1016/j.jfluidstructs.2006.10.008.

Wang, H., Zhu, S., Li, X., Zhang, W., and Nie, Y. (2018a). Numerical simulations of rip currents off arc-shaped coastlines, *Acta Oceanologica Sinica* **37**, 3, pp. 21–30, doi:https://doi.org/10.1007/s13131-018-1197-1.

Wang, J., Ma, Q., and Yan, S. (2018b). A fully nonlinear numerical method for modeling wave–current interactions, *Journal of Computational Physics* **369**, pp. 173–190, doi:https://doi.org/10.1016/j.jcp.2018.04.057.

Wang, J., Ma, Q., and Yan, S. (2021). On extreme waves in directional seas with presence of oblique current, *Applied Ocean Research* **112**, p. 102586, doi:https://doi.org/10.1016/j.apor.2021.102586.

Wang, J. and Ma, Q. W. (2015). Numerical techniques on improving computational efficiency of spectral boundary integral method, *International Journal for Numerical Methods in Engineering* **102**, 10, pp. 1638–1669, doi: https://doi.org/10.1002/nme.4857.

Wang, J., Yan, S., Ma, Q., Wang, J., Xie, Z., and Marran, S. (2020a). Modelling of focused wave interaction with wave energy converter models using qaleFOAM, *Proceedings of the Institution of Civil Engineers - Engineering and Computational Mechanics* **173**, 3, pp. 100–118, doi:https://doi.org/10.1680/jencm.19.00035.

Wang, S., Deng, X., Wang, G., and Yang, X. (2020b). Blending the eddy-viscosity and Reynolds-stress models using uniform high-order discretization. *AIAA Journal* **58**, 12, pp. 5361–5378, doi:https://doi.org/10.2514/1.J059180.

Wang, Z., Yang, J., Koo, B., and Stern, F. (2009). A coupled level set and volume-of-fluid method for sharp interface simulation of plunging breaking waves, *International Journal of Multiphase Flow* **35**, 3, pp. 227–246, doi:https://doi.org/10.1016/j.ijmultiphaseflow.2008.11.004.

Warren, I. R., Larsen, J., and Madsen, P. A. (1985). Application of short wave numerical models to harbour design and future development of the model, in *Proceedings of the International Conference on Numerical and Hydraulic Modelling of Ports and Harbours*, Birmingham.

Watanabe, A., Hara, T., and Horikawa, K. (1984). Study on breaking condition for compound wave trains, *Coastal Engineering in Japan* **27**, 1, pp. 71–82, doi:https://doi.org/10.1080/05785634.1984.11924378.

Watanabe, A. and Maruyama, K. (1986). Numerical modeling of nearshore wave field under combined refraction, diffraction and breaking, *Coastal Engineering in Japan* **29**, 1, pp. 19–39, doi:https://doi.org/10.1080/05785634.1986.11924425.

Watts, P., Grilli, S. T., Kirby, J. T., Fryer, G. J., and Tappin, D. R. (2003). Landslide tsunami case studies using a Boussinesq model and a fully nonlinear tsunami generation model, *Natural Hazards and Earth System Sciences* **3**, 5, pp. 391–402, doi:https://doi.org/10.5194/nhess-3-391-2003.

Waythomas, C. F., Watts, P., Shi, F., and Kirby, J. T. (2009). Pacific basin tsunami hazards associated with mass flows in the Aleutian arc of Alaska, *Quaternary Science Reviews* **28**, 11–12, pp. 1006–1019, doi:https://doi.org/10.1016/j.quascirev.2009.02.019.

Webster, W. C., Duan, W., and Zhao, B. (2011). Green–Naghdi theory, part A: Green–Naghdi (GN) equations for shallow water waves, *Journal of Marine Science and Application* **10**, pp. 253–258, doi:https://doi.org/10.1007/s11804-011-1066-1.

Wei, G., Kirby, J. T., Grilli, S. T., and Subramanya, R. (1995). A fully nonlinear boussinesq model for surface waves. Part 1. Highly nonlinear unsteady waves, *Journal of Fluid Mechanics* **294**, pp. 71–92, doi:https://doi.org/10.1017/S0022112095002813.

Wei, Y., Mao, X.-Z., and Cheung, K. F. (2006). Well-balanced finite-volume model for long-wave runup, *Journal of Waterway, Port, Coastal, and Ocean Engineering* **132**, 2, pp. 114–124, doi:https://doi.org/10.1061/(ASCE)0733-950X(2006)132:2(114).

Weller, H., Tabor, G., Jasak, H., and Fureby, C. (1998). A tensorial approach to computational continuum mechanics using object-oriented techniques, *Computers in Physics* **12**, 6, pp. 620–631, doi:https://doi.org/10.1063/1.168744.

West, B. J., Brueckner, K. A., Janda, R. S., Milder, D. M., and Milton, R. L. (1987). A new numerical method for surface hydrodynamics, *Journal of Geophysical Research: Oceans* **92**, C11, pp. 11803–11824, doi:https://doi.org/10.1029/JC092iC11p11803.

Westerink, J. J., Luettich, R. A., Baptists, A., Scheffner, N. W., and Farrar, P. (1992). Tide and storm surge predictions using finite element model, *Journal of Hydraulic Engineering* **118**, 10, pp. 1373–1390, doi:https://doi.org/10.1061/(ASCE)0733-9429(1992)118:10(1373).

Westhuis, J. H. and Andonowati, A. J. (1998). Applying the finite element method in numerically solving the two dimensional free-surface water wave equations, in *Proceedings of the 13th International Workshop on Water Waves and Floating Bodies*, Hermans, Netherlands, pp. 171–174.

Whitaker, S. (1967). Diffusion and dispersion in porous media, *American Institute of Chemical Engineers Journal* **13**, 3, pp. 420–427, doi:https://doi.org/10.1002/aic.690130308.

Windt, C., Davidson, J., Akram, B., and Ringwood, J. V. (2018). Performance assessment of the overset grid method for numerical wave tank experiments in the OpenFOAM environment, in *Proceedings of the 37th International Conference on Offshore Mechanics and Arctic Engineering*, doi:https://doi.org/10.1115/OMAE2018-77564.

Windt, C., Davidson, J., Schmitt, P., and Ringwood, J. V. (2019). On the assessment of numerical wave makers in CFD simulations, *Journal of Marine Science and Engineering* **7**, 2, p. 47, doi:https://doi.org/10.3390/jmse7020047.

Winterwerp, J. C., De Graaff, R. F., Groeneweg, J., and Luijendijk, A. P. (2007). Modelling of wave damping at Guyana mud coast, *Coastal Engineering* **54**, 3, pp. 249–261, doi:https://doi.org/10.1016/j.coastaleng.2006.08.012.

Woodruff, I., Kirby, J., Shi, F., and Grilli, S. (2018). Estimating meteo-tsunami occurrences for the US east coast, in *Proceedings of the 36th International Coastal Engineering Conference*, p. 66.

Wu, G. (2004). Direct simulation and deterministic prediction of large-scale nonlinear ocean wave-field PhD thesis, Massachusetts Institute of Technology, USA.

Wu, G. and Eatock-Taylor, R. (1995). Time stepping solutions of the two-dimensional nonlinear wave radiation problem, *Ocean Engineering* **22**, 8, pp. 785–798, doi:https://doi.org/10.1016/0029-8018(95)00014-C.

Wu, G. X. and Eatock-Taylor, R. (1994). Finite element analysis of two-dimensional non-linear transient water waves, *Applied Ocean Research* **16**, 6, pp. 363–372, doi:https://doi.org/10.1016/0141-1187(94)00029-8.

Wu, G. X. and Hu, Z. Z. (2004). Simulation of nonlinear interactions between waves and floating bodies through a finite-element-based numerical tank, *Proceedings of the Royal Society of London. Series A: Mathematical, Physical and Engineering Sciences* **460**, 2050, pp. 2797–2817, doi:https://doi.org/10.1098/rspa.2004.1302.

Wu, G. X., Ma, Q. W., and Eatock-Taylor, R. (1995). Nonlinear wave loading on a floating body, in *Proceedings of the 10th International Workshop on Water Waves and Floating Bodies*, Oxford, UK.

Wu, G. X., Ma, Q. W., and Eatock-Taylor, R. (1996). Analysis of interactions between nonlinear waves and bodies by domain decomposition, in *Proceedings of the ONR 21st Symposium on Naval Hydrodynamics*, Trondheim, Norway, pp. 110–119.

Wu, Y. and Cheung, K. F. (2008). Explicit solution to the exact Riemann problem and application in nonlinear shallow-water equations, *International Journal for Numerical Methods in Fluids* **57**, 11, pp. 1649–1668, doi:https://doi.org/10.1002/fld.1696.

Xiang, J., Latham, J.-P., Vire, A., Anastasaki, E., and Pain, C. C. (2012). Coupled Fluidity/Y3D technology and simulation tools for numerical breakwater modelling, *Proceedings of the International Conference on Coastal Engineering, ICCE 2012* **1**, 33, pp. 57–66, doi:https://doi.org/10.9753/icce.v33.structures.66.

Xiao, W., Liu, Y., Wu, G., and Yue, D. K. P. (2013). Rogue wave occurrence and dynamics by direct simulations of nonlinear wave-field evolution, *Journal of Fluid Mechanics* **720**, pp. 357–392, doi:https://doi.org/10.1017/jfm.2013.37.

Xing, Y. (2017). Numerical methods for the nonlinear shallow water equations, in R. Abgrall, C.-W. Shu (ed.), *Handbook of Numerical Analysis*, Vol. 18 (Elsevier, Amsterdam, NL), pp. 361–384.

Xu, H., Cantwell, C. D., Monteserin, C., Eskilsson, C., Engsig-Karup, A. P., and Sherwin, S. J. (2018). Spectral/hp element methods: Recent developments, applications, and perspectives, *Journal of Hydrodynamics* **30**, pp. 1–22, doi:https://doi.org/10.1007/s42241-018-0001-1.

Xu, H. and Yue, D. K. (1992). Computations of fully nonlinear three-dimensional water waves, in *Proceedings of the 19th Symposium on Naval Hydrodynamics*, Seoul, Korea.

Xue, M., Xü, H., Liu, Y., and Yue, D. K. P. (2001). Computations of fully nonlinear three-dimensional wave–wave and wave–body interactions. Part 1. Dynamics of steep three-dimensional waves, *Journal of Fluid Mechanics* **438**, pp. 11–39, doi:https://doi.org/10.1017/S0022112001004396.

Yan, S. (2006). Numerical simulation on nonlinear response of moored floating structures to steep waves, PhD thesis, City University London, UK.

Yan, S. and Ma, Q. W. (2010). QALE-FEM for modelling 3D overturning waves, *International Journal for Numerical Methods in Fluids* **63**, 6, pp. 743–768, doi:https://doi.org/10.1002/fld.2100.

Yan, S. and Ma, Q. W. (2011). Improved model for air pressure due to wind on 2D freak waves in finite depth, *European Journal of Mechanics – B/Fluids* **30**, 1, pp. 1–11, doi:https://doi.org/10.1016/j.euromechflu.2010.09.005.

Yan, S. and Ma, Q. W. (2014). Sensitivity investigation on wave dynamics with thin-walled moonpool, in *Proceedings of the 24th International Ocean and Polar Engineering Conference*, Busan, Korea.

Yan, S., Ma, Q. W., and Adcock, T. A. A. (2010). Investigation of freak waves on uniform current, in *Proceedings of the 25th International Workshop on Water Waves and Floating Bodies*, Harbin, China.

Yan, S., Zhou, J. T., Ma, Q. W., Wang, J., Zheng, Y., and Wazni, B. (2013). Fully nonlinear simulation of tsunami wave impacts on onshore structures, in *Proceedings of the 23rd International Ocean and Polar Engineering Conference*, Anchorage, Alaska.

Yang, Z. and Liu, P. L.-F. (2020). Depth-integrated wave–current models. Part 1. Two-dimensional formulation and applications, *Journal of Fluid Mechanics* **883**, doi:https://doi.org/10.1017/jfm.2019.831.

Yang, Z. and Liu, P. L.-F. (2022). Depth-integrated wave–current models. Part 2. Current with an arbitrary profile, *Journal of Fluid Mechanics* **936**, doi: https://doi.org/10.1017/jfm.2022.42.

Young, I. R. and Babanin, A. V. (2006). Spectral distribution of energy dissipation of wind-generated waves due to dominant wave breaking, *Journal of Physical Oceanography* **36**, 3, pp. 376–394, doi:https://doi.org/10.1175/JPO2859.1.

Youngs, D. L. (1982). Time-dependent multi-material flow with large fluid distortion. K. W. Morton and M. J. Baines (eds.), *Numerical Methods for Fluid Dynamics*, Academic Press, London, UK.

Yu, T. L. and Guan, C. L. (2000). Numerical study with finite element method for combined wave model of refraction-diffraction-dissipation due to wave breaking (in Chinese), *Journal of Ocean University of Qingdao* **30**, 01, pp. 15–21.

Yu, X., Isobe, M., and Watanabe, A. (1992). Finite element solution of wave field around structures in nearshore zone, *Coastal Engineering in Japan* **35**, 1, pp. 21–33, doi:https://doi.org/10.1080/05785634.1992.11924555.

Yu, X. and Togashi, H. (1995). Irregular waves over an elliptic shoal, in *Proceedings of the 24th International Conference on Coastal Engineering*, pp. 746–760, doi:https://doi.org/10.1061/9780784400890.056.

Zakharov, V. E. (1968). Stability of periodic waves of finite amplitude on the surface of a deep fluid, *Journal of Applied Mechanics and Technical Physics* **9**, 2, pp. 190–194, doi:https://doi.org/10.1007/BF00913182.

Zelt, J. (1991). The run-up of nonbreaking and breaking solitary waves, *Coastal Engineering* **15**, 3, pp. 205–246, doi:https://doi.org/10.1016/0378-3839(91)90003-Y.

Zhang, J.-X., Sukhodolov, A. N., and Liu, H. (2014a). Non-hydrostatic versus hydrostatic modelings of free surface flows, *Journal of Hydrodynamics* **26**, 4, pp. 512–522, doi:https://doi.org/10.1016/S1001-6058(14)60058-5.

Zhang, L. and Edge, B. L. (1997). A uniform mild-slope model for waves over varying bottom, in *Proceedings of the 25th International Conference on Coastal Engineering*, pp. 941–854, doi:https://doi.org/10.1061/9780784402429.074.

Zhang, L. B. (1996). A modified hybrid element model for combined diffraction-refraction-reflection-dissipation waves over large regions, *Chinese Journal of Oceanology and Limnology* **14**, 1, pp. 68–78, doi:https://doi.org/10.1007/BF02850543.

Zhang, R. and Stive, M. J. (2019). Numerical modelling of hydrodynamics of permeable pile groins using SWASH, *Coastal Engineering* **153**, p. 103558, doi:https://doi.org/10.1016/j.coastaleng.2019.103558.

Zhang, T., Huang, Y. J., Liang, L., Fan, C. M., and Li, P. W. (2018). Numerical solutions of mild slope equation by generalized finite difference method, *Engineering Analysis with Boundary Elements* **88**, pp. 1–13, doi:https://doi.org/10.1016/j.enganabound.2017.12.005.

Zhang, T., Lin, T., Lin, C., and Huang, Y. J. (2021). Numerical simulation of extended mild-slope equation including wave breaking effect, *Engineering Analysis with Boundary Elements* **128**, pp. 42–57, doi:https://doi.org/10.1016/j.enganabound.2021.03.018.

Zhang, X. (2018). Multi-model method for simulating 2D surface-piercing wave-structure interactions, PhD thesis, City, University of London.

Zhang, Y. and Baptista, A. M. (2008). SELFE: A semi-implicit Eulerian–Lagrangian finite-element model for cross-scale ocean circulation, *Ocean Modelling* **21**, 3-4, pp. 71–96, doi:https://doi.org/10.1016/j.ocemod.2007.11.005.

Zhang, Y., Kennedy, A. B., Donahue, A. S., Westerink, J. J., Panda, N., and Dawson, C. (2014b). Rotational surf zone modeling for $O(\mu^4)$ Boussinesq–Green–Naghdi systems, *Ocean Modelling* **79**, pp. 43–53, doi:https://doi.org/10.1016/j.ocemod.2014.04.001.

Zhang, Y., Kennedy, A. B., Panda, N., Dawson, C., and Westerink, J. J. (2013). Boussinesq–Green–Naghdi rotational water wave theory, *Coastal Engineering* **73**, pp. 13–27, doi:https://doi.org/10.1016/j.coastaleng.2012.09.005.

Zhang, Y., Kennedy, A. B., Tomiczek, T., Donahue, A., and Westerink, J. J. (2016a). Validation of Boussinesq–Green–Naghdi modeling for surf zone hydrodynamics, *Ocean Engineering* **111**, pp. 299–309, doi:https://doi.org/10.1016/j.oceaneng.2015.11.004.

Zhang, Y. J., Ye, F., Stanev, E. V., and Grashorn, S. (2016b). Seamless cross-scale modeling with SCHISM, *Ocean Modelling* **102**, pp. 64–81, doi:https://doi.org/10.1016/j.ocemod.2016.05.002.

Zhao, B. B., Duan, W. Y., and Ertekin, R. C. (2014). Application of higher-level GN theory to some wave transformation problems, *Coastal Engineering* **83**, pp. 177–189, doi:https://doi.org/10.1016/j.coastaleng.2013.10.010.

Zhao, B. B., Duan, W. Y., Ertekin, R. C., and Hayatdavoodi, M. (2015). High-level Green–Naghdi wave models for nonlinear wave transformation in three dimensions, *Journal of Ocean Engineering and Marine Energy* **1**, pp. 121–132, doi:https://doi.org/10.1007/s40722-014-0009-8.

Zhao, E., Sun, J., Tang, Y., Mu, L., and Jiang, H. (2020). Numerical investigation of tsunami wave impacts on different coastal bridge decks using immersed boundary method, *Ocean Engineering* **201**, p. 107132, doi:https://doi.org/10.1016/j.oceaneng.2020.107132.

Zhao, L., Panchang, V., Chen, W., Demirbilek, Z., and Chhabbra, N. (2001). Simulation of wave breaking effects in two-dimensional elliptic harbor wave models, *Coastal Engineering* **42**, 4, pp. 359–373, doi:https://doi.org/10.1016/S0378-3839(00)00069-7.

Zhao, Y. and Anastasiou, K. (1993). Economical random wave propagation modelling taking into account non-linear amplitude dispersion, *Coastal Engineering* **20**, 1–2, pp. 59–83, doi:https://doi.org/10.1016/0378-3839(93)90055-D.

Zhao, Z., Huang, P., Li, Y., and Li, J. (2013). A lattice Boltzmann method for viscous free surface waves in two dimensions, *International Journal for Numerical Methods in Fluids* **71**, 2, pp. 223–248, doi:https://doi.org/10.1002/fld.3660.

Zheng, J., Li, R. J., and Jiang, S. H. (2013). Application of a nonlinear elliptic mild slope equation in Rizhao harbor, *Applied Mechanics and Materials* **405**, doi:https://doi.org/10.4028/www.scientific.net/AMM.405-408.1449.

Zheng, Y. H., Shen, Y. M., and Qiu, D. H. (2001). Application of nonlinear dispersion relation in solving hyperbolic mild slope equations (in Chinese), *Journal of Hydraulic Engineering* **2**, pp. 69–75.

Zhou, J. G. (2002). A lattice Boltzmann model for the shallow water equations, *Computer Methods in Applied Mechanics and Engineering* **191**, 32, pp. 3527–3539, doi:https://doi.org/10.1016/S0045-7825(02)00291-8.

Zhou, J. G., Causon, D. M., Mingham, C. G., and Ingram, D. M. (2001). The surface gradient method for the treatment of source terms in the shallow-water equations, *Journal of Computational Physics* **168**, 1, pp. 1–25, doi: https://doi.org/10.1006/jcph.2000.6670.

Zienkiewicz, O., Taylor, R., and Zhu, J. (2013). *The Finite Element Method: Its Basis and Fundamentals*, 7th edn. (Butterworth-Heinemann, Oxford), doi:https://doi.org/10.1016/C2009-0-24909-9.

Zijlema, M. (2010). Computation of wind-wave spectra in coastal waters with SWAN on unstructured grids, *Coastal Engineering* **57**, 3, pp. 267–277, doi: https://doi.org/10.1016/j.coastaleng.2009.10.011.

Zijlema, M. (2012). Modelling wave transformation across a fringing reef using SWASH, in *Proceedings of the 33rd International Conference on Coastal Engineering, ICCE 2012*, Santander, Spain, doi:https://doi.org/10.9753/icce.v33.currents.26.

Zijlema, M., Stelling, G., and Smit, P. (2011). SWASH: An operational public domain code for simulating wave fields and rapidly varied flows in coastal waters, *Coastal Engineering* **58**, 10, pp. 992–1012, doi:https://doi.org/10.1016/j.coastaleng.2011.05.015.

Zijlema, M. and Stelling, G. S. (2005). Further experiences with computing non-hydrostatic free-surface flows involving water waves, *International Journal for Numerical Methods in Fluids* **48**, 2, pp. 169–197, doi:https://doi.org/10.1002/fld.821.

Zijlema, M. and Stelling, G. S. (2008). Efficient computation of surf zone waves using the nonlinear shallow water equations with non-hydrostatic pressure, *Coastal Engineering* **55**, 10, pp. 780–790, doi:https://doi.org/10.1016/j.coastaleng.2008.02.020.

Zijlema, M., Van Vledder, G. P., and Holthuijsen, L. (2012). Bottom friction and wind drag for wave models, *Coastal Engineering* **65**, pp. 19–26, doi: https://doi.org/10.1016/j.coastaleng.2012.03.002.

Zou, Q. H., Hongchuan, W., and Junning, P. (1997). Wave-current propagation over a frictional topography, *Journal of Nanjing Hydraulic Research Institute* **4**.

Zuo, Q. H., Yao, G. Q., and Ding, B. C. (1993). Refraction and diffraction of waves caused by topography of friction drag (in Chinese). *China Harbour Engineering* **01**, pp. 34–40.

Zubier, K., Panchang, V., and Demirbilek, Z. (2003). Simulation of waves at Duck (North Carolina) using two numerical models, *Coastal Engineering Journal* **45**, 3, pp. 439–469, doi:https://doi.org/10.1142/S0578563403000853.

Index

P

parabolic MSE (PMSE), 30, 35–38
phase-averaged models, 9, 121–122, 129
phase-resolving models, 3–4, 6, 24, 27, 68, 121–122, 124–125, 129–130, 132
pollutant transport, 90, 131–132
poro-elasticity, 108–109
porosity, 92–93
porous coastal structures, 91, 105
potential flow, 4, 7, 27, 48–49, 58–59, 63, 65, 67, 69, 71–73, 101, 109, 126, 130, 133
potential flow models, 4, 7, 27, 63, 65, 67, 69, 71, 73
pressure gradient, 38, 52, 81, 98
(wave) propagation, 1, 3, 9, 13–14, 27, 29–32, 34–39, 41–43, 46, 48–49, 51–53, 55, 57, 68–72, 87–88, 90, 99, 106, 114, 126, 129, 131

R

radiation stress, 19, 25, 124
random waves, 37
Rayleigh equation, 54
Reef3D, 105, 133
(wave) reflection, 3, 14, 24, 29, 34, 36, 68, 102, 125, 127, 130
(wave) refraction, 3, 14, 24, 29–32, 34–35, 37, 123, 130
regular waves, 32, 34–35, 37, 103, 110, 114, 126
relaxation zones, 102, 121, 123, 126–128
resonance theories, 15
Reynolds stress model, 100
Reynolds-averaged Navier–Stokes (RANS), 75, 91–92
roughness, 16, 22
run-up, 1, 39, 41–43, 50, 52, 88–89, 114, 123–124

S

Saint-Venant equations, 38
scales, 1–4, 23, 29, 45, 49, 52, 80, 83–84, 98–100, 111, 129–131
SCHISM, 44–45, 82, 133
scour, 104–105, 108, 131–132
sediment, 3, 6, 50, 87, 89–90, 107–108, 114–115, 117, 131–132
sediment transport, 6, 50, 87, 89, 107–108, 114, 117, 132
shallow water, 4, 18, 21, 23–25, 29, 37–39, 42, 45, 47, 50, 53, 57, 70, 78, 80–81, 84, 88, 113–114, 124, 126, 130–131, 133
shallow water equations, 24, 38–39, 45, 113–114, 126
shear stress, 22, 40, 80
sheared current, 24–25, 34, 54–55, 61, 68, 70, 130
sheltering theory, 15
ship hydrodynamics, 106–107
(wave) shoaling, 3, 13–14, 21, 24, 29, 32, 50, 56, 130
SLOSH, 44
smooth particle hydrodynamics (SPH), 103
solid mechanics, 6, 107, 117
solitary waves, 43, 54, 57, 69, 87, 103, 126
spectral element method, 70–71
spectral forcings, 5
spectral models, 3, 11, 14–15, 21–22, 24, 65, 12–130, 133
(wave) spectrum, 11, 16–22, 37, 65–66, 122–123
storm surge, 1–2, 38–39, 43–45, 77, 122–123, 130
stream function, 63–64
super-harmonics, 18–19, 25
surf, 1, 21, 35, 37, 57, 88, 103–104, 122–123, 126, 133
SWAN, 17, 20, 23, 123–124, 133
SWASH, 1, 59, 84, 87–89, 103, 133
swell, 1, 22, 25